Environmental Science and Engineering

Series Editors

The ultimate goal of this series is to contribute to the protection of our environment, which calls for both profound research and the ongoing development of solutions and measurements by experts in the field. Accordingly, the series promotes not only a deeper understanding of environmental processes and the evaluation of management strategies, but also design and technology aimed at improving environmental quality. Books focusing on the former are published in the subseries Environmental Science, those focusing on the latter in the subseries Environmental Engineering.

Chaolei Yuan
Editor

Proceedings of 6th International Conference on Resources and Environmental Research - ICRER 2024

Springer

Editor
Chaolei Yuan
School of Agriculture and Biotechnology
Sun Yat-sen University
Shenzhen, China

ISSN 1863-5520 ISSN 1863-5539 (electronic)
Environmental Science and Engineering
ISBN 978-3-031-93294-6 ISBN 978-3-031-93295-3 (eBook)
https://doi.org/10.1007/978-3-031-93295-3

This Springer imprint is published by the registered company Springer Nature Switzerland AG
The registered company address is: Gewerbestrasse 11, 6330 Cham, Switzerland

Preface

The 6th International Conference on Resources and Environmental Research (ICRER 2024) was held successfully in Hong Kong, China, from December 6th to 8th, 2024. This event was co-organized by the University of Hong Kong and the Beijing CAS Industrial Energy and Environment Technology Institute, with technical support from Xiamen University of Technology, the University of Nottingham Ningbo, Qatar University, and some other institutions.

ICRER 2024 attracted approximately 50 submissions, with around 35 attendees from various countries, including China, Japan, the United States, Indonesia, Thailand, and Malaysia. Researchers from diverse fields presented their findings, contributing to the vibrant discussions and knowledge exchange that defined the event. We extend our sincere gratitude to all participants whose involvement made the conference a success.

The program featured two distinguished keynote speakers and three invited speakers. In addition to oral presentations, both in-person and virtual, poster presentations were also showcased. Each submitted paper underwent a rigorous peer review process, and the accepted works are included in this volume, which is divided into five thematic sections: 1) Climate Change and Meteorological Extremes, 2) Environmental Pollution and Ecological Risk Assessment, 3) Hydrological Dynamics and Soil Conservation Strategies, 4) Sustainable Resource Management and Circular Economy, and 5) Environmental Engineering Technologies and Applications.

We would like to extend our heartfelt thanks to the organizing committee, technical committee, and volunteers whose hard work and dedication ensured the smooth running of the event. Special appreciation is also extended to all session chairs for their invaluable contributions to the success of the conference.

Chaolei Yuan
ICRER 2024 Publication Chair

Committees

Conference Chair

Kaimin Shih	The University of Hong Kong, China

Program Chairs

Dennis Y. C. Leung	The University of Hong Kong, China
Chuyang Y. Tang	The University of Hong Kong, China
Xueming Chen	Fuzhou University, China
Miwako Hosoda Seisa	University, Tokyo, Japan

Publication Chair

Chaolei Yuan	Sun Yat-sen University, China

Local Chairs

Zhao Qi	The University of Hong Kong, China
Tan Giin Yu Amy	City University of Hong Kong, China

Steering Committees

Mohamed El-Din	University of Alberta, Canada
Satya Harpalani	Southern Illinois University Carbondale, USA

Technical Committee

Xueyan Zhang	Dalian University of Technology, China
Lingchen Mao	University of Shanghai for Science and Technology, China
Xishan Li	Dalian Maritime University, China
Kuichang Zuo	Peking University, China
Zhen Hu	Shandong University, China

Contents

Climate Change and Meteorological Extremes

Spatiotemporal Variations of Landfalling Tropical Cyclones in Mainland China Intensity and ENSO

Nan Wang(✉), Zitong Zhao, Ziyi Gao, Ke Bu, Ningyi Zhang, and Hongyue Gu

Department of Resources and Environment, Faculty of Science, Jiamusi University, Jiamusi 154007, China
562856392@qq.com

Abstract. Based on the tropical cyclone data of the Joint Typhoon Warning Center (JTWC), this study explores the spatial and temporal variation characteristics of tropical cyclone activity in China during the 70 years from 1951 to 2020, and the possible spatial and temporal variation mechanism of ENSO on landing tropical cyclone activity. The main conclusions are as follows: (1) the tropical Accumulated Cyclone Energy index (ACE) in China has obvious characteristics of inter-annual and chronological changes. From the long-term change trend, the cyclone energy index is significantly weakened. (2) No significant correlation was found between the cyclone energy index and the sea surface temperature in the Niño 3.4 zone. The correlation coefficient of the cyclone energy index and the sea surface temperature in ENSO 3.4 area was positive from 1951 to the late 1990s, negative from the early 2000s to 2020, and mutations occurred around 1967 and 1994. (3) The area with high observation intensity is located in the southeast area of Taiwan Island (120–130°E, 15–25°N), and the second high value area is the South China Sea (110–115°E, 15–20°N). The spatial distribution of the first eigenvector of the observed intensity EOF is essentially the same as that for many years. The time coefficient of the first mode changed twice in 1972 and 1982, indicating that the trend weakened after 1972, and the trend strengthened rapidly by 1982. The spatial distribution of the second eigenvector is mainly found in the northwest Pacific Ocean (130–140°E, 15–25°N) region, and the negative center is mainly distributed near the archipelago of the Philippine group (120–130°E, 10–20°N), showing a positive-negative spatial distribution feature. The time coefficient of the second mode indicates that such spatial distribution features often show opposite trends, and the time coefficient is mutated around 1977, indicating that the opposite trend of observed intensity began to weaken in the late 1977s. In the year of ElNiño, the observation intensity in the south China coast of South China and the northeastern part of the Japanese Sea of the Philippines was increasing, with the decrease mainly distributed in the Bohai Sea, Yellow Sea, East China Sea, the Korean Peninsula and the eastern and western waters of the Philippines. In the year of LaNiña, the observation intensity of coastal areas in China decreased, while the trend of observation intensity increased in the Philippines.

Keywords: tropical cyclone · ACE · ENSO

C. Yuan (Ed.): ICRER 2024, ESE, pp. 3–18, 2025.
https://doi.org/10.1007/978-3-031-93295-3_1

1 Introduction

China is one of the countries most seriously affected by typhoons in the world. It not only has many typhoons landed in China, according to statistics since 1951, about 7–8 typhoons have landed in the southeast coast of China every year, and the typhoon intensity is large, causing serious economic losses and casualties. Especially in recent decades, with the rapid economic development of China's coastal areas and the population is more dense, the disaster bearing capacity of these areas gradually declines, and the property losses and casualties caused by tropical cyclones are more serious (Pielke et al. 1998; 2008). Similarly, in China, the 1956 super typhoon "dai" in Zhejiang province, only thousands of deaths in Zhejiang, in addition, 7503 in 1975 typhoon "Nina", brought huge precipitation to inland Henan province, although limited to historical political reasons lack of exact disaster statistics, according to estimates because of "75.8" flood of secondary disasters affected by tens of thousands of people. On July 31 and August 1, 1996, typhoon "herb" made landfall in Taiwan and Fujian respectively, with landfall intensity was strong typhoon and strong tropical storm respectively. The intensity of "herb" is not particularly strong, but the circulation is very large, the water vapor is particularly sufficient, and the late path deep inland, so cause a large range of strong wind and heavy rain in East China, North China, Taiwan Alishan 24 h rainfall reached 1748 mm, the North China Plain local process rainfall exceeded 600 mm, more than 700 people died and 110,000 people were injured. Since 1983, an average of 505 deaths and direct economic losses per year have amounted to RMB 37 billion yuan in 2008 (Zhang et al. 2011).

Some scholars have noticed the interannual and chronological changes of monsoon circulation, ocean circulation, ENSO and North Atlantic Oscillation (NAO), and believe that the changes of these factors may be the main reason for the interchronological oscillation or even longer scale climate cycle of typhoon activity. For example, Gray (1999) believes that the activity of the Atlantic hurricane is closely related to the strength of warm salt circulation: warm salt circulation strengthens, North Atlantic SST warming, convection strengthens, hurricane activity strengthens; on the contrary, warm salt circulation weakens, North Atlantic SST cooling, convection weakens, and hurricane activity weakens. He also believed that the long-term change of warm salt circulation has a time scale strength change cycle of about 20–50a. However, those who support the view that tropical cyclone activity is affected by global warming argue that changes in warm-salt circulation can itself be affected by global warming. For example, the analysis of He et al. (1999) showed that the frequency of typhoons in the Northwest Pacific Ocean decreased in ElNiño, and the frequency of typhoons in LaNiña increased, and they believed that different phases in ENSO correspond to different atmospheric circulation modes, thus affecting the change of typhoon frequency.

The reason for the difference in the impact of ENSO on northwest Pacific typhoons may be due to the differences in statistical methods and the different data used. The research of Wu et al. (2003) and Lin and Zhang (2004) based on CMA typhoon Yearbook data showed that during the El Niño period, the frequency of strong typhoons in the Northwest Pacific Ocean decreased (especially in the next year of El Niño), and the frequency of typhoons affecting China decreased, but the intensity increased, while El Niño reversed. Wu (2007) analyzed that the weak wind shear and warm SST facilitate the generation of more storms, and the Atlantic hurricane season starts earlier and ends later, which leads to the increase of PDI; the source and path of Sahara atmosphere (SAL) and tropical cyclones have an important impact on the change of hurricane intensity over the last 30 years. Other studies have shown that there is a clear link between the ENSO and the Atlantic and Pacific Ocean (Knutson 2006). It is a negative correlation, that is, El Niño, in the Atlantic and western Pacific Australia, the number of typhoons decreased. Since the intensity and frequency of El Niño events have increased over the past 30 years (i.e., Pacific Decade Oscillation (PDO) phase) and the warming phase of ENSO, the tropical cyclone trend may be influenced by the combination of ENSO and global warming.

In the context of global warming, changes in tropical cyclone activity, such as numbers may increase, life histories may increase, changes in intensity may occur, and can be predicted through climate patterns. However, there are still many problems, such as the global or regional occurrence frequency, geographical distribution, path change, landing frequency and intensity in coastal areas, there is still a lot of uncertainty, and it is still difficult to produce a definite result.

2 Data and Methodology

2.1 Sources of Date

Tropical cyclone data were obtained from the United States Typhoon Warning Center (JTWC) (https://metoc.ndbc.noaa.gov/en/web/guest/jtwc/products-and-services-notice) 1951–2020 Tropical cyclone best path data set. Sea surface temperature (SST) observations from 1951–2020 were obtained from the Met Office Hadley Centre. (http://hadobs.metoffice.com/hadisst/data /download.html) $1° \times 1°$ global monthly sea surface temperature-sea ice density combined observation data HadSST1 (Rayner et al. 2003).The abnormal sea surface temperature in May-November of Niño-3.4 was obtained from the reconstructed sea surface temperature data extended by the Climate Prediction Center (http://www.cpc.noaa.gov/data/indices/) ERSSTv4 (Huang et al. 2015; Liu et al. 2015; Smith and Reynolds 2003; Smith and Reynolds 2004; Smith et al. 2008; Xue et al. 2003).

2.2 Research Method

(1) EOF empirical orthogonal function analysis method

EOF empirical orthogonal function analysis method was used to analyze the spatial and temporal change of energy accumulation, and the t-test method was used to test the significance of the mean difference of the conclusion.

Principle and algorithm:

Select the data to be analyzed and do data preprocessing, usually in a flat form. To obtain a data matrix $X_{m\times n}$.

The cross-product of X and its transpose X^T matrix is calculated to obtain the square matrix:

$$C_{m\times n} = \frac{1}{n} X \times X^T$$

If X is already processed to be flat, C is called a covariance matrix; if X has been standardized (i.e., the average of data for each row in C is 0 and the standard deviation is 1), C is called a correlation coefficient matrix.

Calculate the characteristic root $(\lambda_{1,\ldots,m})$ of the square matrix C and eigenvectors $V_{m\times n}$, both meet:

$$C_{m\times n} \times V_{m\times n} = V_{m\times n} \times \Lambda_{m\times n}$$

Among them: Λ is the m × n dimensional diagonal array, i.e.:

$$\Lambda = \begin{bmatrix} \lambda_1 & 0 & \cdots & 0 \\ 0 & \lambda_2 & \cdots & 0 \\ \cdots & \cdots & \cdots & \cdots \\ 0 & 0 & \cdots & \lambda_m \end{bmatrix}$$

In general, the feature root is arranged in order from large to small, i.e., $\lambda_1 > \lambda_2 > L > \lambda_m$, Since the data X is the true observed value, the ¸ should be greater than or equal to 0. Each feature root of a non-zero corresponds to a list of eigenvector values, also called EOF. For example, the eigenvector value corresponding to 1 is the first EOF mode, which is the first column of V, i.e., $EOF_1 = (:, 1)$; The eigenvector corresponding to k is the k-th column of V, i.e., $EOF_k = (:, k)$. Calculate principal components. Proting EOF onto the raw data matrix X yields the time coefficient (i.e., principal components) corresponding to all spatial eigenvectors, i.e.:

$$PC_{m\times n} = V^T_{m\times n} \times X_{m\times n}$$

where each row of data in the PC is the time coefficient corresponding to each eigenvector. The first line of PC (1, :) is the time coefficient of the first EOF, and so on.

Above are the EOF and principal components (PC) calculated from the data matrix X, so the original data matrix X can also be completely recovered using EOF and PC, i.e.:

$$X = EOF \times PC$$

(2) Accumulated Cyclone Energy (ACE)

Bell et al. proposed the concept of Accumulated Cyclone Energy (ACE), which comprehensively considered the life period and the maximum average wind speed of a tropical cyclone, reflecting the energy accumulation of tropical cyclones. The longer the life period, the greater the wind speed, the more energy the tropical cyclone accumulated, and the greater its potential destructive force. Bell defines the cyclone energy index (ACE) as the cumulative square sum of maximum wind speed (m/s) every 6 h near the tropical cyclone near a tropical storm (17.2 m s^{-1}) or above, using the cyclone energy index to represent the destruction potential of a tropical cyclone.

$$\mathrm{ACE} = \sum_{i=1}^{m}\sum_{i=1}^{n} v_j^2$$

i represents the frequency of landing of a tropical cyclones during a year; j represents the number of counts every 6 h for any tropical cyclone, and v_j represents the maximum average wind speed near the center of the tropical cyclone every 6 h.

3 Long-Term Trend of Accumulated Cyclone Energy (ACE)

From 1951 to 2020, the cyclone energy index reached its historical maximum in 1958 and 7.06×10^8 $(m/s)^2$ was the minimum and only 1.17×10^8 $(m/s)^2$ in 1978, with an average value of 3.08×10^8 $(m/s)^2$. In Fig. 1a, it shows the rapid enhancement period from the early 1950s to the late 1950s, a significant decline and recovery trend in the late 1960s to the early 2000s, and a rapid decline period from the late 2000s to 2020 (red fitting curve). In terms of the long-term trend, the cyclone energy index weakened and passed the significant level test of 0.01 (P = 0.0029). However, Emanuel (2005) showed that the PDI index increased in the western Pacific and Atlantic Ocean, which is largely related to the time scale of the tropical cyclones used.

The sliding T-test of the cyclone energy index time series can be seen in Fig. 1b that the T statistics in 1972 and 1984 exceeded the significant level test (T0.01 = 2.878) indicates that the cyclone energy index is likely to change twice in 1972 and 1984, the first from the high value period before 1972 to the low value period, and the second from the low value period before 1987 to the high value period.

After landing, the cyclone energy index reached the historical maximum in 1958 in 2.40×10^4 $(m/s)^2$ and the minimum in 1.30×10^3 $(m/s)^2$ in 1967, and the average value was 4.35×10^3 $(m/s)^2$. In Fig. 1c, showing a similar trend to that in Fig. 1a, and the correlation coefficient between cyclone parameter index and cyclone energy index after landing is 0.55 (through the significant level test of 0.01). From the long-term trend, the cyclone energy index weakened after landing (failed the significant level test of 0.05).

By sliding T-test on the time series of cyclonic energy index after landing, it can be seen in Fig. 1d that the significant level test exceeded 0.05 in 1964 and 1988 (T0.05 = 2.101), which shows that the cyclone energy index is likely to change twice in 1964 and 1988, the first time changed from the high value period before 1964 to the low value period, and the second time from the low value period before 1988 to the high value period before 1988.

(a) time series of the cyclone energy index 10^6 $(m/s)^2$; (b) sliding T curve of the cyclone energy index; (c) time series of the cyclone energy index after landing 10^6

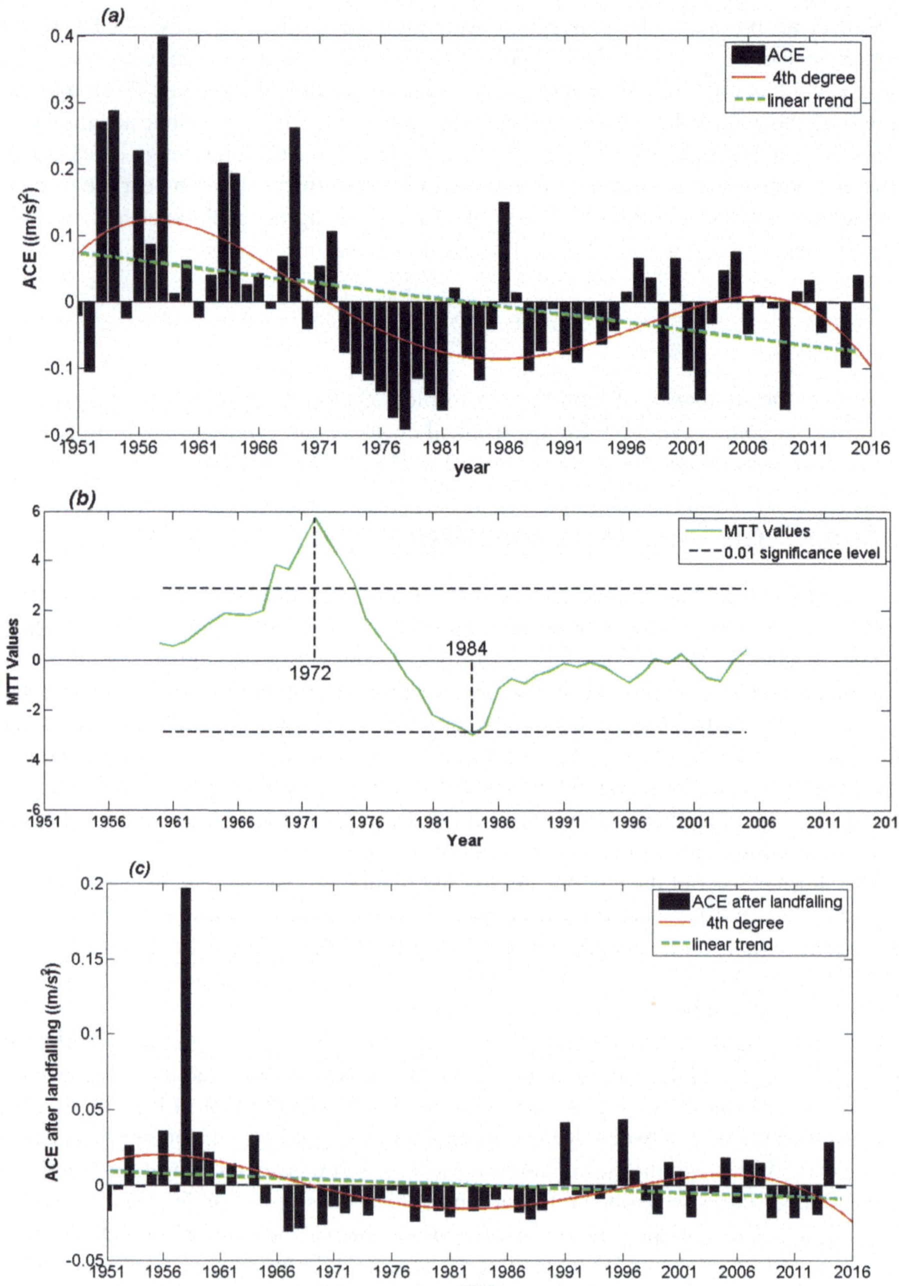

Fig. 1. Time change of landing tropical cyclone energy index in 1951–2020.

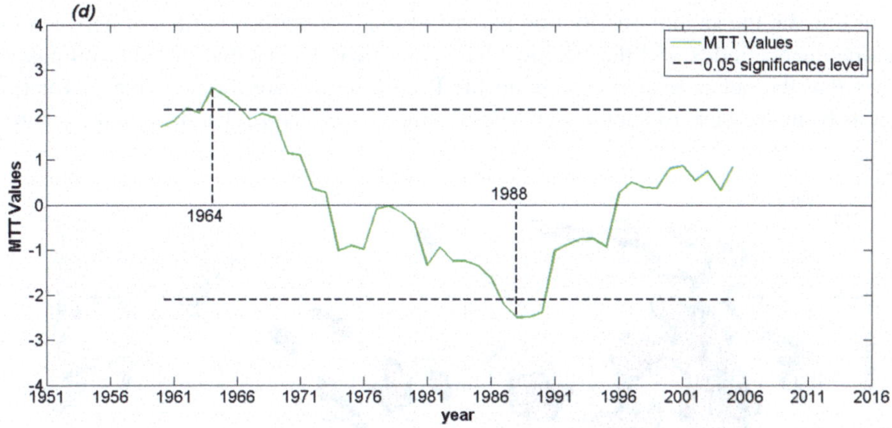

Fig. 1. (*continued*)

(m/s)^2; (d) sliding T curve of the cyclone energy index after landing. The black dashed line refers to the 0.01 significant level reliability line in Fig. 1b, and the black dashed line refers to the 0.05 significant level reliability line in Fig. 1d.

4 Statistical Relationship Between the Accumulated Cyclone Energy (ACE) and the ENSO

In Fig. 2, from 1951 to the late 1990s, the cyclone energy index followed the Niño 3.4 index, but showed the opposite trend from the early 2000s to 2020. There is a significant positive correlation between cyclone energy index and Niño 3.4 index ($p = 0.14$), and the correlation coefficient is $R = 0.18$.

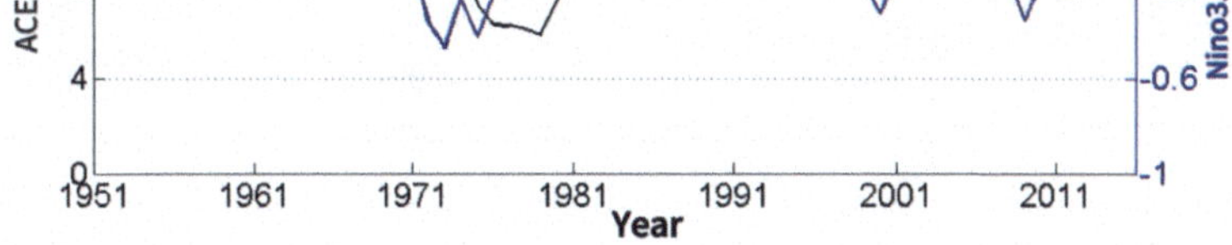

Fig. 2. The 5-year sliding mean curve of the cyclone energy index (10^{-4} (m/s)^2) and the SSTA index in the Niño 3.4 area.

In Fig. 3a, the spatial distribution pattern of the correlation coefficient between the cyclone energy index and the Pacific SST is clarified. The spatial distribution pattern shows that the better related regions in the Pacific region are the negative correlation regions from the western Pacific warm pool to the western North Pacific and the positive

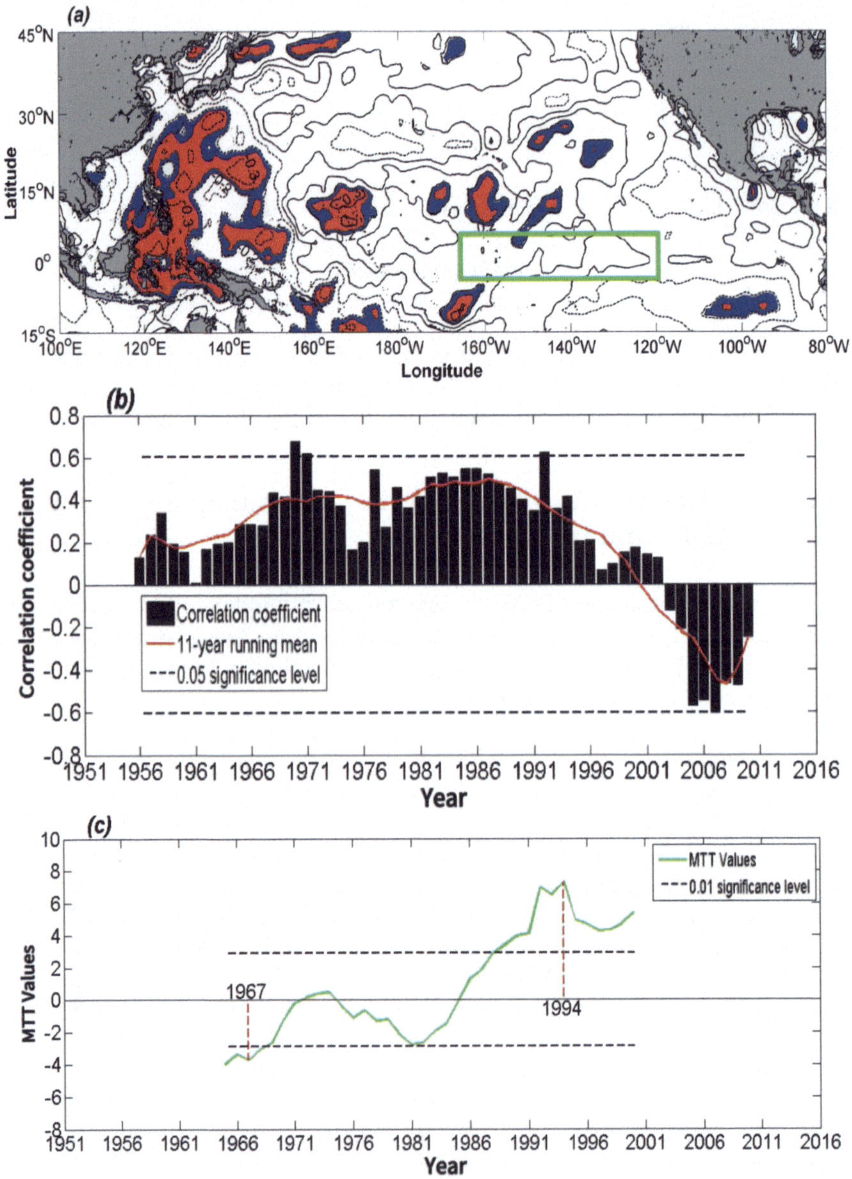

Fig. 3. Spatial distribution of correlation coefficients between ACE and Pacific SST and sliding correlation coefficients with SSTA index in Niño 3.4 region time series, T test.

correlation regions of the equatorial Middle East Pacific and the northern and eastern sea areas of the North Pacific. Significant negative correlations were distributed in the South China Sea, the eastern Philippines, the West Pacific Pool (120–160°E, 0–30°N), and significant positive correlations were scattered in the Pacific Ocean, the northwest Pacific (145–160°E, 20–30°N) and the central, northern and eastern parts of the North Pacific.

No significant correlation between the cyclone energy index and the sea surface temperature was found in the Niño 3.4 area (green rectangular zone).

(a) spatial distribution of correlation coefficient between ACE and Pacific SST; (b) time series of 11-year sliding correlation coefficient between ACE and SSTA index in Niño 3.4 area; (c) sliding T test. In Fig. 3a, the red (light blue) shade indicates over the 95% (90%) confidence level, and the green rectangle indicates the position of the Niño 3.4 area.

The next step analyzed the interannual variation of the sliding correlation coefficient of sea surface temperature in Niño 3.4 and cyclone energy index (Fig. 3b). The sliding correlation coefficient was positive from 1951 to the late 1990s, which passed the significant level test of 0.05 in 1970 and 1971. The sliding correlation coefficient was negative from the early 2000s to 2020, with the significant level test of 0.05 passed in 2007. The stage is very obvious. The sliding T-test shows in Figs. 4, 5, 6 and 7c that the statistics of the 1967 and 1994 T values exceeded the significant level test of 0.01 ($T0.01 = 2.878$). It can be determined that the correlation between the two occurred around 1967 and 1994. Before 1967 was a weaker positive correlation, then changed to a stronger positive correlation. In 1994, the strong transition process occurred, which was a strong positive correlation before, and then the rapid transition was a strong negative correlation later.

5 Spatiotemporal Variation Characteristics of Tropical Cyclone Observation Intensity

The intensity of tropical cyclone observation is the average of the near-center maximum wind speed recorded by all tropical cyclones entering the $5° \times 5°$longitude grid cell from May to November of the year. The observed intensity can partly reflect the extent to which a region is affected by a tropical cyclone.

Figure 4 depicts the spatial distribution pattern of the observation intensity in 1951–2020 (May-November), indicating that the high value area of the observed intensity is located in the southeast waters of Taiwan Island (120–130°E, 15–25°N), and the sub-high value area is the South China Sea (110–115°E, 15–20°N).

From 1951 to 2020, the first mode variance of the observed intensity EOF expansion contributed 64.10%, and the variance of the second mode contributed 5.64%.

From the spatial distribution of the first eigenvector of the observed intensity EOF (Fig. 5a), both eigenvectors are positive in the critical region. It is mainly located in the southeastern waters of Taiwan Island (120–130°E, 15–25°N) and the South China Sea (110–115°E, 15–20°N). The spatial distribution is basically the same as that of many years (Fig. 4). The essence of the spatial distribution of the first eigenvector of the observed intensity EOF is the spatial distribution characteristic of the observed intensity

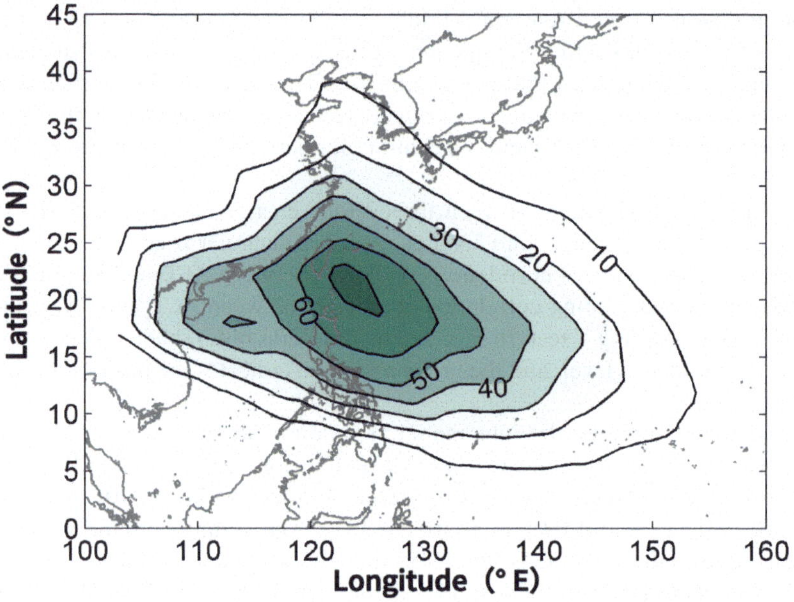

Fig. 4. Spatial distribution characteristics of tropical cyclone observation intensities (m/s).

of tropical cyclones. The sliding T-test of the time coefficient of the first mode found that two changes occurred in 1972 and 1982 (Fig. 5b), indicating that the centralized distribution trend weakened after 1972, and the centralized distribution trend began to strengthen after 1982.

From the spatial distribution of the second mode (Fig. 5c), the positive value center mainly appears in the northwest Pacific (130–140°E, 15–25°N) region, and the negative value center is mainly distributed near the archipelago of the Philippine group (120–130°E, 10–20°N), showing positive-negative spatial distribution. The time coefficient of the second mode indicates that the observed intensity of the tropical cyclone in the northwest Pacific (130-140°E, 15-25°N) often shows opposite trends near the archipelago of the Philippine Group. The sliding T-test of the time coefficient of the second mode found a mutation around 1977, indicating that the observed intensity of the tropical cyclone (130–140°E, 15–25°N) in the northwest Pacific Ocean and the archipelago of the Philippine group often began to weaken in the late 1977s (Fig. 5d).

(a) the spatial distribution of the first mode; (b) the temporal coefficient of the first mode; (c) the spatial distribution of the second mode; and (d) the temporal coefficient of the second mode.

6 Relation Between Spatial Variation of Tropical Cyclone Observation Intensity and ENSO

The intensity of tropical cyclone observation is the average of the near-center maximum wind speed recorded by all tropical cyclones entering the 5° × 5° longitude grid cell from May to November of the year. The observed intensity can partly reflect the extent to which a region is affected by a tropical cyclone.

Figure 6a shows the spatial distribution pattern of the average annual observation intensity in normal years, indicating that the high intensity area is located in the eastern waters of Taiwan Island, and the sub-high value area is the South China Sea area. Figure 6b shows the comparison between ElNiño years and normal years. The high positive areas mainly appear in the areas 140–160°E, 5–22°N, as well as the coast of South China and the northeast of the Sea of Japan, indicating that the intensity of these areas is increasing compared with the normal years, reflecting the increasing threat of these areas

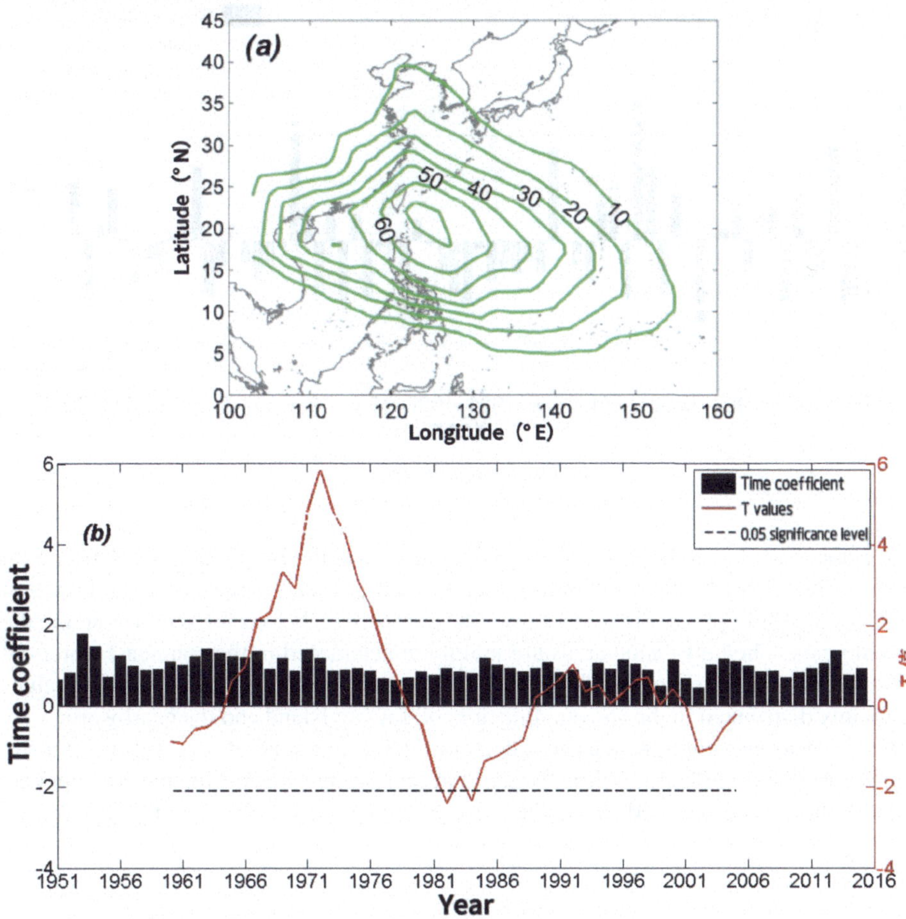

Fig. 5. EOF decomposition of tropical cyclone observed intensities.

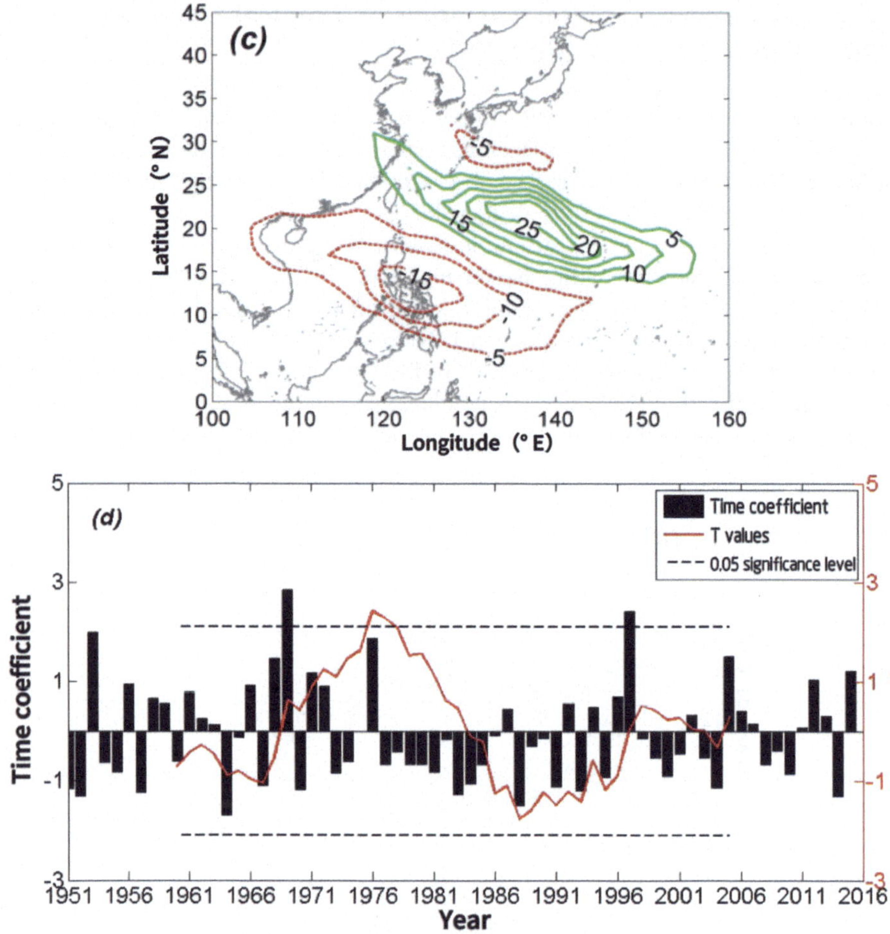

Fig. 5. (*continued*)

from tropical cyclones. Negative areas mainly appear in the Bohai Sea, the Yellow Sea, the East China Sea, the Korean Peninsula, and the other area is in the eastern and western waters of the Philippines. Figure 6c shows the comparison between LaNiña year and the normal years. The high-value areas are mainly concentrated in the southeast waters of Japan, the eastern waters of the Philippines and the South China Sea. The negative values are mainly distributed in the surrounding area of Taiwan Island and the nearby area with 140ºE, 15ºN as the center, which can reflect the decreasing trend of observation intensity in the coastal areas of China, while the trend of increasing observation intensity appears in the Philippines. Figure 6d shows the comparison between LaNiña and ElNiño years, with the high value appearing in the East China Sea and the east ocean, the Philippines and surrounding waters. Negative values occur in the eastern part of Taiwan island and in the nearby area centered on 145°E, 20°N. It reflects the greater threat in China's Bohai Rim, East China and the Philippines, and the weakening threat across the Taiwan Strait and southern China.

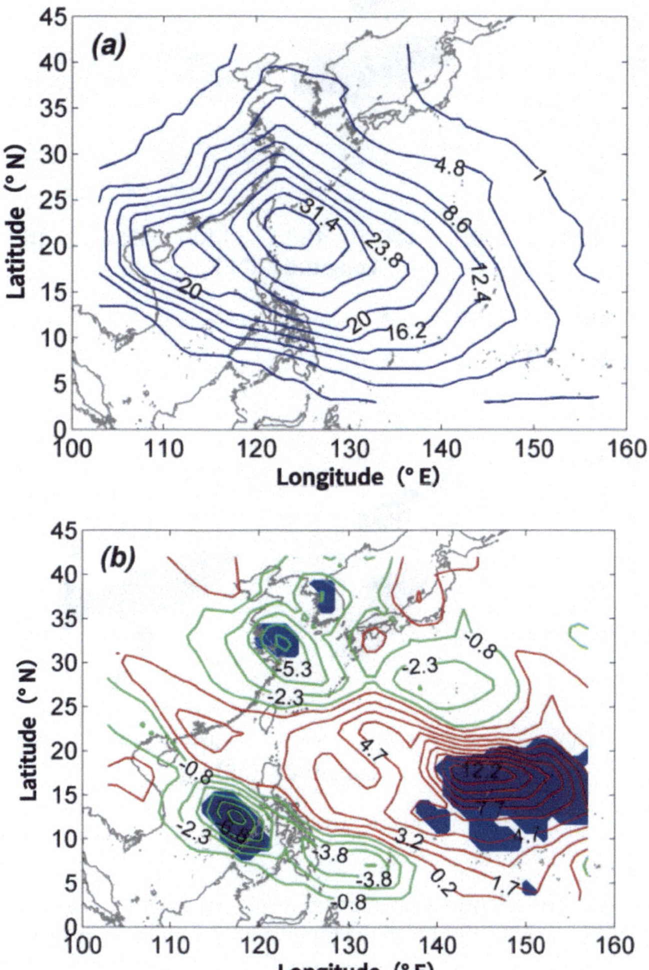

Fig. 6. Geographic distribution of average annual tropical cyclone observation intensity (m/s) within the 5° × 5° grid.

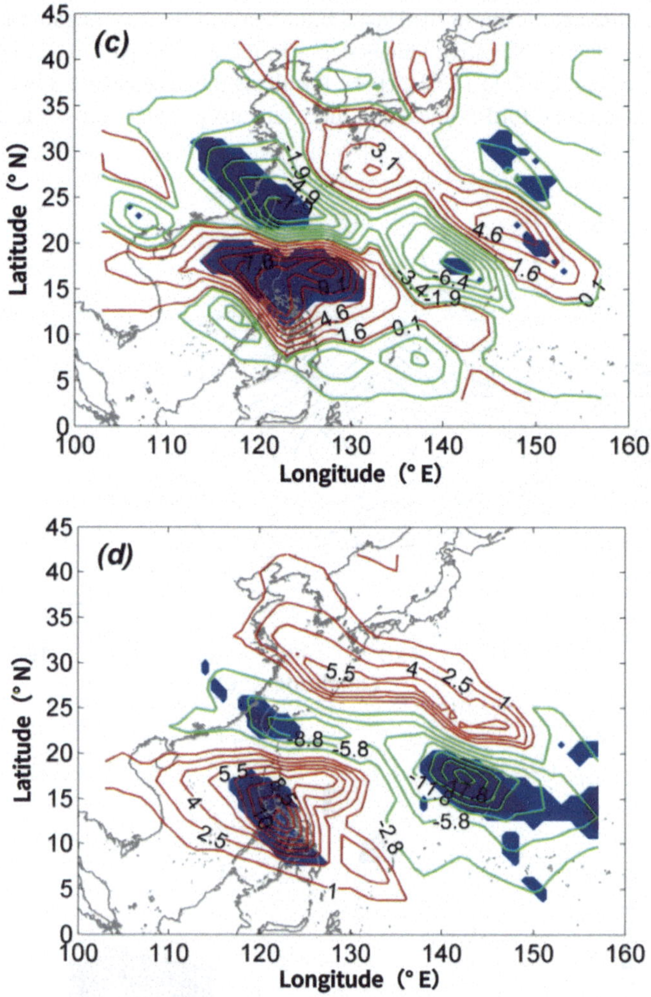

Fig. 6. (*continued*)

(a) normal year; (b) ElNiño year-normal year; (c) LaNiña year-normal year; (d) LaNiña year-ElNiño year. The light (dark) blue shaded area in the plot indicates that the mean difference significance test passed the 90% (95%) t-test.

7 Conclusions

(1) Accumulated Cyclone Energy (ACE) showed a period of rapid enhancement from the early 1950s to the late 1950s, a trend of significant decline from the late 1960s to the early 2000s, and a period of rapid decline from the late 2000s to 2020. The cyclone energy index changed twice in 1972 and 1984, the first period changed from the high value period before 1972 to the low value period, and the second period

before 1987. From the long-term change trend, the cyclone energy index showed a significant weakening trend, and the cyclone energy index also showed a weakening trend after landing.

(2) No significant correlation was found of the cyclone energy index with the SST in the Niño 3.4 region. From the 1951 to the late 1990s, they were positive and negative from the early 2000s to 2020, passing the significant level test of 0.05 in 2007. The correlation was mutated around 1967 and 1994. Before 1967 was a weaker positive correlation, then changed to a stronger positive correlation. In 1994, the strong transition process occurred, which was a strong positive correlation before, and then the rapid transition was a strong negative correlation later.

(3) The observed high value area is located in the southeastern waters of Taiwan Island (120–130°E, 15–25°N), and the sub-high value area is the South China Sea (110–115°E, 15–20°N).

The spatial distribution of the first eigenvector of the observed intensity EOF is essentially the same. The time coefficient of the first mode changed twice in 1972 and 1982, indicating that the trend weakened after 1972, and the trend was rapidly strengthened by 1982. From the perspective of the spatial distribution of the second mode, the positive value center mainly appears in the northwest Pacific Ocean (130–140°E, 15–25°N) region, and the negative value center is mainly distributed near the archipelago of the Philippine group (120–130°E, 10–20°N), with a positive-negative spatial distribution feature. The time coefficient of the second mode indicates that such spatial distribution features often show opposite trends, and the time coefficient is mutated around 1977, indicating that the opposite trend of observed intensity began to weaken in the late 1977s.

The observed intensity of tropical cyclones when ElNiño landed in China tended to increase in South China coast, northeast of the Sea of Japan, east and west waters of the Philippines, the reduction is mainly distributed in the Bohai Sea, the Yellow Sea, the East China Sea, the Korean Peninsula and the eastern and western waters of the Philippines. In LaNiña, the observation intensity of coastal areas in China decreased, while the trend of observation intensity increased in the Philippines.

Acknowledgements. The Innovation and Entrepreneurship Project for College Students in Heilongjiang Province (S202410222064). This work was supported by the Jiamusi University Doctoral Special Research Fund Project (JMSUBZ2019-11). Heilongjiang Provincial Universities Basic Scientific Research Project (2019-KYYWF-1399). The authors are grateful to the editor and anonymous reviewers for their helpful comments and also acknowledge helpful comments by Chenglin Gu.

References

1. Pielke, R.A.J., Landsea, C.W.: Normalized hurricane damages in the United States: 1925–95. Weather Forecast. **13**(3), 621–631 (2008)
2. Pielke, R.A.J., Gratz, J., Landsea, C.W., et al.: Normalized hurricane damage in the United States: 1900–2005. Nat. Hazard. Rev. **9**(1), 29–42 (2008)
3. Wen, K.: Chinese Meteorological Disasters Pandect. Science Press, Beijing (2008). [Winkerang: China Meteorological Disasters Grand Ceremony (Comprehensive Volume). Meteorological Publishing House, Beijing (2008)]

4. Zhang, J., Wu, L., Zhang, Q.: Trend analysis of tropical cyclone disasters in China under the background of global warming. **27**(4), 442–454 (2011)
5. Gray, W.M.: On the Causes of Multidecadal Climate Change and Prospects for Increased Atlantic Basin Hurricane Activity in the Coming Decades, pp. 378–381. American Meteorological Society, Boston (1999)
6. He, M., Song, W., Chen, X.: El Nino and anti-El Nino events with northwest Pacific typhoon activity. J. Trop. Meteorol. **15**(1), 17–25 (1999)
7. Wu, D., Bai, Y., Zhang, H.: Effects of tropical cyclones from water temperature changes in the subsurface layers of the equatorial Western Pacific warm pool. J. Trop. Meteorol. **19**(3), 253–259 (2003)
8. Lin, H., Zhang, Y.: The climatic characteristics affecting tropical cyclone activity in China and their relationship with Pacific SST. Trop. P **20**(2), 218–224 (2004)
9. Wu, L.G., Wang, B., Scott, A.B.: Implications of tropical cy clone power dissipation index. Int. J. Climatol. (in press). Published online in Wiley Inter Science (www.interscience.wiley.com). https://doi.org/10.1002/joc.2007,1573
10. Knutson T. Possible Relationships Between Climate Change and Tropical Cyclone Activity. http://citeseerx.ist.psu.edu/viewdoc/download?doi=10.1.1.177.6640&rep=rep1&type=pdf
11. Rayner, N.A., Parker, D.E., Horton, E.B.: Global analyses of sea surface temperature, sea ice, and night marine air temperature since the late nineteenth century. J. Geophys. Res. **108**(D14), 4407 (2003)
12. Liu, K., Chan, J.C.: Climatological characteristics and seasonal forecasting of tropical cyclones making landfall along the south China coast. Mon. Weather Rev. **131**(8), 1650–1662 (2003)
13. Huang, B., Banzon, V.F., Freeman, E., et a1.: Extended reconstructed sea surface temperature version 4 (ERSST.v4). Part I: upgrades and intercomparisons. J. Clim. **28**(3), 911–930 (2015)
14. Smith, T.M., Reynolds, R.W.: Extended reconstruction of global sea surface temperatures based on COADS data (1854–1997)[J]. J. Clim. **16**(16), 1495–1510 (2003)
15. Smith, T.M., Reynolds, R.W.: Improved extended reconstruction of SST (1854–1997). J. Clim. **17**(17), 2466–2477 (2004)
16. Smith, T.M., Reynolds, R.W., Thomas, C., et al.: Improvements to NOAA's historical merged land-ocean surface temperature analysis (1880–2006). J. Clim. **21**, 2283–2296 (2008)
17. Xue, Y., Smith, T.M., et al.: Interdecadal changes of 30-Yr SST normals during 1871–2000. J. Clim. **16**(10), 1601–1612 (2003)
18. Bell, G.D., Halpert, M.S., Schnell, R.C., et al.: Climate assessment for 1999. Bull. Am. Meteorol. Soc. **81**(6), S1–S50 (2000)
19. Trenberth, K.E.: Uncertainty in hurricanes and global warming. Science **308**(5729), 1753–1754 (2005)
20. Song, J.J., Wang, Y., Wu, L.: Trend discrepancies among three best track data sets of western north Pacific tropical cyclones. J. Geophys. Res. Atmos. **115**, D12128 (2010)

Characteristics of a Snowstorm Event in Changsha of China Based on Multi-source Observations

Mei Hu[1,2], Zhanman Deng[3](✉), Lei Shu[4], Jiaowen Huang[1,2], and Tiebin Yin[5]

[1] Hunan Meteorological Observatory, Changsha 410118, China
[2] Hunan Key Laboratory of Meteorological Disaster Prevention and Reduction, Changsha 410118, China
[3] Hunan Meteorological Disaster Prevention Technology Center, Changsha 410053, China
728764596@qq.com
[4] Zhuzhou Meteorological Bureau, Zhuzhou 412000, China
[5] Lianyuan Meteorological Bureau, Lianyuan 417100, China

Abstract. The characteristics of a snowstorm event in Changsha of China in 2021 are analyzed based on multi-source observations. Results show that before the snowstorm, the dry layer disappeared, the moist layer thickened, and the wet-bulb temperature was below 0 °C. The saturated stratification in the cold wedge below the inversion layer was conducive to the aggregation growth of falling snowflakes. Analysis of the wind profile radar data shows a close correlation between the intermittent light rain (snow) in Changsha on December 25 and the passage of several short-wave troughs. The temporary weakening of snowfall on December 26 is attributed to the transition from southwesterly wind to westerly wind. Besides, the correlation coefficient of 0.9–0.97 and uniform differential reflectivity are the characteristics of snowfall echoes. The analysis of the laser raindrop disdrometer data reveals that in the rain stage, particles had high falling velocity and small diameter, and the droplet spectrum was narrow. In the sleet stage, the droplet spectrum broadened rapidly, with an increase in particle number and diameter. In the snow stage, particles had lower falling velocity and larger diameter, and the droplet spectrum was broader. Additionally, the average droplet spectrum distribution (single peak pattern) of all stages is very close to that in the snow stage. The number concentration of raindrops decreases rapidly as the particle diameter increases. Overall, the multi-source new-type observation data with high spatio-temporal resolutions are effective in phased precipitation forecasts.

Keywords: L-band radiosonde radar · Wind profile radar · S-band dual-polarization radar · Laser raindrop disdrometer · Precipitation phase

1 Introduction

Snowstorms are one of the hazardous weather events in central China in winter. They often have substantial impacts on transportation, electricity, communication and construction [1, 2], and are thus concerned by the governments and public. Past studies

C. Yuan (Ed.): ICRER 2024, ESE, pp. 19–32, 2025.
https://doi.org/10.1007/978-3-031-93295-3_2

on snowstorm mainly focused on synoptic diagnostic analysis. They comprehensively considered temperature, thickness and other indicators at different levels, and summarized the key points for predicting snowstorms [3–6]. However, due to the relatively complex physical mechanisms involved in the rain-snow transition and the cessation of nighttime manual observations in China, forecasting precipitation phase and snowfall amount remains a focal point and challenge in heavy snowfall prediction.

The widespread application of advanced detection instruments in recent years, including the dual-polarization Doppler weather radar, wind profiler and laser raindrop disdrometer, has improved the spatiotemporal resolution in detecting hazardous weather events [7, 8]. On this basis, the evolution, dynamics and physical mechanisms of hazardous weather events have been better observed and understood, and the hazardous-weather monitoring and forecasting ability has been improved [9, 10]. According to Cheng et al. [11], the dual-polarization Doppler weather radar, compared with the conventional single-polarization Doppler radar, has higher spatiotemporal resolution, and can measure differential reflectivity factor (Z_{DR}), specific differential phase (K_{DP}) and correlation coefficient (CC). It promotes the real-time observation and classification of precipitation particles, and may provide more detailed precipitation-phase features such as ice crystal and hail. Zhou et al. [12] demonstrated that the wind profiler radar can detect cold air intrusion earlier than traditional weather maps. In addition, the wind profiler radar can be used to analyze weather systems such as pressure troughs and ridges, as well as cold and warm advection. Li et al. [13] and Zhou et al. [14] pointed out the raindrop disdrometer data can be used to analyze the microphysical characteristics of precipitation particles during rainfall and snowfall events. Li et al. [13] and Zhou et al. [14] also highlighted the importance of continuous 1-min interval observation by laser raindrop disdrometers in capturing various types of precipitation. These observations can compensate for the limitations in conventional radiosonde and radar detection.

Multi-source observations are important in the characterization of precipitation evolution and weather monitoring and warning [15, 16], but there are few studies on snowstorms by using multi-source observations. Moreover, the variation law of precipitation phase varies from place to place due to the difference of temperature, terrain, latitude and other factors. The application of multi-source observations is of great significance to improve the operational snowfall forecasting in China [17]. In this study, a snowstorm process in Changsha City, Hunan Province of China from December 25 to 26, 2021 is selected as an example. By using the L-band radiosonde radar, wind profiler radar, S-band dual-polarization radar, laser raindrop disdrometer and the intensive surface observations, we explore the effect of multi-source new observation data in precipitation phase forecast in China. We hope our findings may provide scientific reference for the study of snowfall weather based on multi-source observations.

2 Data and Synoptic Situation

2.1 Data

The new-type observation data used in this study are from multiple sources, namely the SA dual-polarization weather radar, the L-band radiosonde radar, the wind profiler radar, and the laser raindrop disdrometer. The conventional surface observation data is from national meteorological observation stations. All the aforementioned observational instruments are located in Lianhua Town, Yuelu District, Changsha City (112.7°E, 28.1°N).

2.2 Synoptic Situation of the Snowstorm Process

The surface air temperature dropped below 0 °C from 16:00 UTC, December 25 in Changsha station (Fig. 1). Changsha experienced rain before 17:00 UTC, sleet from 17:00 UTC to 18:00 UTC and intermittent light snow during 18:00–20:00 UTC. After 20:00 UTC, precipitation phase transitioned into continuous snowfall, which lasted for 22 h with a total snow amount of 17.2 mm, and the snowfall amount reached the level of snowstorm.

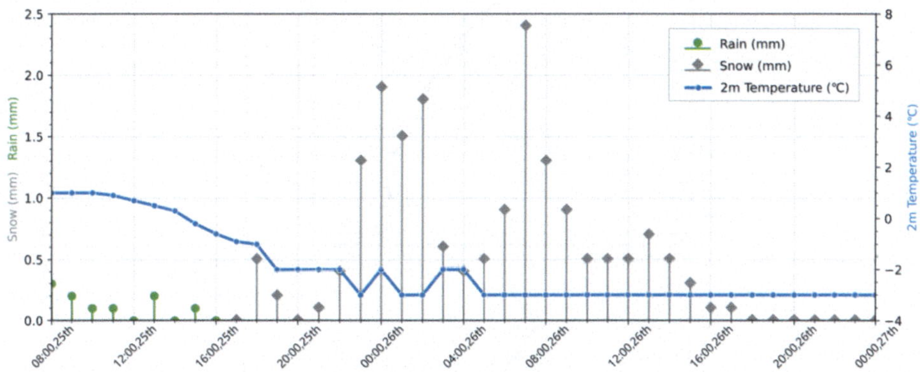

Fig. 1. The variations of hourly precipitation (mm) and 2 m temperature from 12:00 UTC December 24 to 00:00 UTC December 27, 2021.

3 Analysis of the L-Band Radiosonde Radar

To investigate the influence of the atmospheric conditions on the precipitation phase transition, we use the radiosonde profiles (Fig. 2). At 12:00 UTC on 25 December 25 (Fig. 2a), northerly wind prevailed below 925 hPa. A pronounced dry layer was observed at the height between −10 °C and −20 °C, which hindered the occurrence of snowfall. At 00:00 UTC on December 26 (Fig. 2b), the wind at 500 hPa and 700 hPa shifted from westerly to southwesterly. It led to the disappearance of the dry layer. The entire atmosphere was in a near-saturated stratification. The moist layer, with relative humidity exceeding 80%, extended up to 350 hPa (8.3 km) where the wet-bulb temperature was −35.6 °C (Table 1). These conditions are highly conducive to the formation of ice crystals and snowflakes. Besides, the southeasterly wind at 850 hPa turned to northeasterly wind. The northerly wind below 925 hPa reached 12 m s^{-1}, indicating the thickening of lower-level cold wedge. The wet-bulb temperature of the whole layer was below 0 °C and the surface temperature dropped to −2.5 °C (Table 1), providing extremely favorable temperature conditions for snowstorm. A weak inversion layer was observed between 700 hPa and 1000 hPa. This layer can increase the snowflake aggregation when the snowfall falling downward, thus strengthening snowfall. Additionally, due to the existence of the inversion layer, the water vapor mixing ratio above the inversion layer was high under saturated conditions. Thus, the strong southwesterly wind may transport abundant water vapor above the inversion layer. These conditions can lead to the occurrence of heavy snowfall.

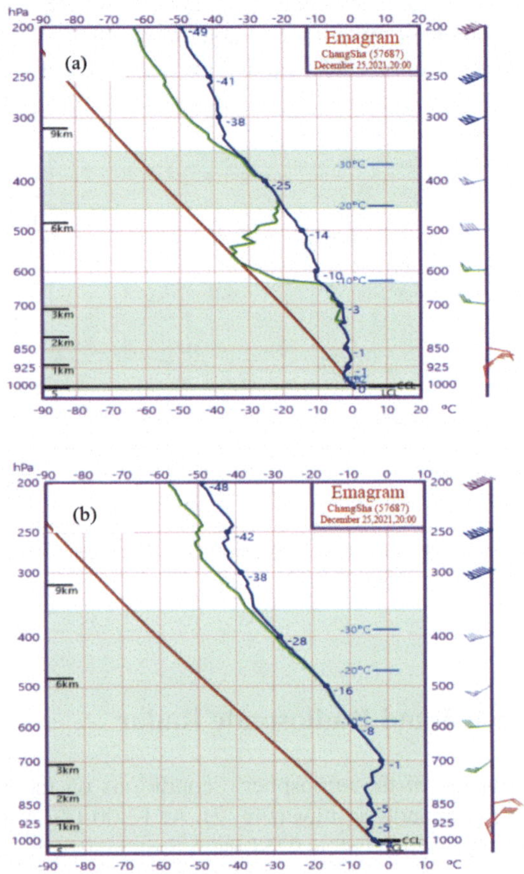

Fig. 2. The T-lnP diagram at Changsha station (56778) at (a) 12:00 UTC on December 25 and (b) 00:00 UTC on December 26, 2021.

Table 1. Radiosonde data of wet bulb temperature and relative humidity at Changsha station at 12:00 UTC on December 25 and at 00:00 UTC on December 26, 2021.

Variables	Wet bulb temperature (°C)						Relative humidity (%)					
Levels (hPa)	1000	925	850	700	500	350	1000	925	850	700	500	350
12:00 UTC, December 25	−0.3	−1.4	−1.8	−4	−17.7	−33.1	100	100	100	91	27	90
00:00 UTC, December 26	−5.0	−4.9	−5.0	−1.8	−16.4	−35.6	100	100	100	100	96	80

4 Analysis of the Wind Profiler Radar

Figure 3 illustrates the time-height profile of horizontal wind obtained from the wind profiler radar in Changsha station. From 12:00 UTC December 25 to 18:00 UTC December 26 (solid black line in Fig. 3a), there were three periods of "northwesterly wind shifting to southwesterly wind" (short curve in Fig. 3a). It indicates that shortwave trough passed, which can enhance the dynamic lifting condition over the station and results in intermittent light rainfall (snowfall) on December 25. From 19:30 UTC to 20:00 UTC, December 25, northwesterly wind prevailed within 2.5–3.5 km, and it shifted to southwesterly wind from 20:00 UTC (the red triangle in Fig. 3a). At that time, persistent snowfall started. Above 2.5 km, the southwesterly wind extended vertically (the black dashed arrows in Fig. 3a). At the altitudes exceeding 4.3 km, the southwesterly wind reached above 20 m s^{-1}. With the strengthening of the warm-moist southwesterly jet above the cold wedge, the dynamical conditions were further intensified. Then, Changsha experienced its first snowfall peak, with 3-h snowfall amount reaching 5.2 mm. From 02:00 UTC to 05:00 UTC, December 26, the snowfall weakened. During this period, within 3.0–4.5 km (the black square in Fig. 3b) the southwesterly wind shifted to westerly wind. Around 03:30 UTC, it reverted to southwesterly wind again. It is hypothesized that the temporary snowfall amount decrease during this period is related to the horizontal wind variation within 3.0–4.5 km (characterized by the shift to westerly wind). Around 04:00 UTC December 26 (the red triangle in Fig. 3b), the southeasterly wind below 2.0 km shifted to northeasterly wind (the black rectangle in Fig. 3b). The prevailing northeasterly airflow within the cold wedge was conducive to the upward movement of warm-moist airflow over the cold wedge. At that time, the second snowfall peak occurred in Changsha, with 3 h snowfall of 4.6 mm and hourly snowfall amount of 2.4 mm. After 12:00 UTC on December 26 (figure omitted), the northwesterly wind started to prevail above 2.5 km, and easterly wind prevailed below 2.5 km. The snowfall gradually weakened and eventually ceased. In general, the spatio-temporal distribution of horizontal wind from the wind profiler radar well reflects the variation characteristics of horizontal wind during the snowfall process.

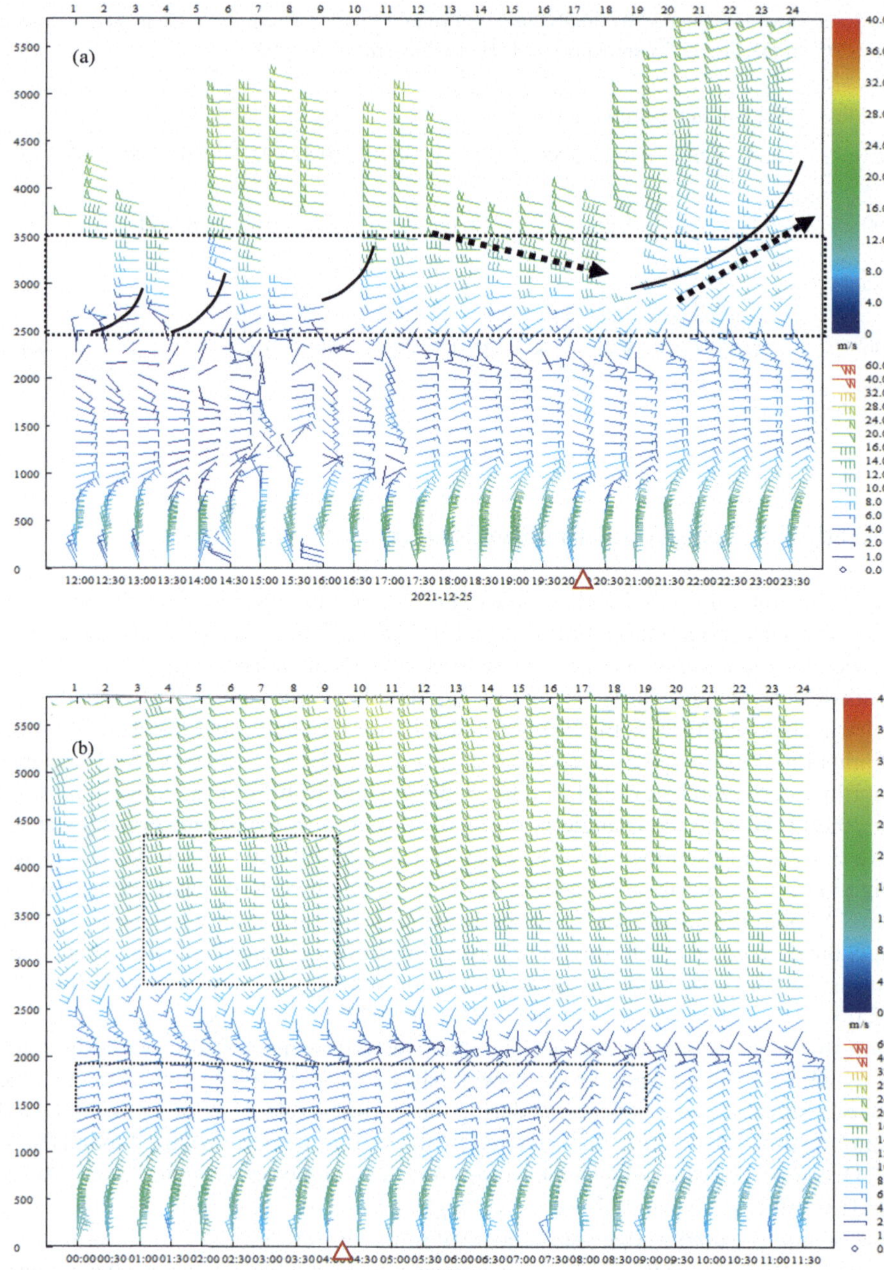

Fig. 3. (a) Horizontal wind time-height profile from the wind profile radar at Changsha station from 12:00 UTC December 25 to 00:00 UTC December 26, 2021 (unit: m s^{-1}). The x-axis is time (UTC). Red triangles indicate the onset of persistent snowfall. (b) The same as Fig. 3a except for the period from 00:00 UTC December 26 to 12:00 UTC December 26, 2021.

5 Analysis of the Dual-Polarization Radar

5.1 Analysis in the Rainfall Period with Dual-Polarization Radar Data

Figure 4a shows the distribution of Z_h (horizontal reflectivity factor) values at 0.5° elevation angle at 17:52 UTC December 25. Wide areas with Z_h greater than 15 dBz were observed, with the highest value reaching 35–45 dBz. As can be seen, higher Z_h values (40–45 dBz) correspond to higher Z_{DR} values (1–2 dB), and lower Z_h values (15–25 dBz) correspond to lower Z_{DR} values (0.2–1 dB) in Fig. 4b. A vertical profile (Fig. 4c) has been obtained along the white line in Fig. 4a, revealing high Z_h values (40–45 dBz) in 0.5–1.5 km. The region highlighted by the double red line in Fig. 4c represents a horizontal bright band with Z_h greater than 30 dBz, indicating the bright band of the melting layer [18]. Correspondingly, in 0.5–1.5 km there is a low CC band, as indicated by the double white lines in Fig. 4d, with the CC values ranging from 0.8 to 0.9. Due to the inferior water vapor conditions and ascent motion in winter, precipitation events in winter fail to generate large particles. The mixture of particles in different phases leads to a lower CC value.

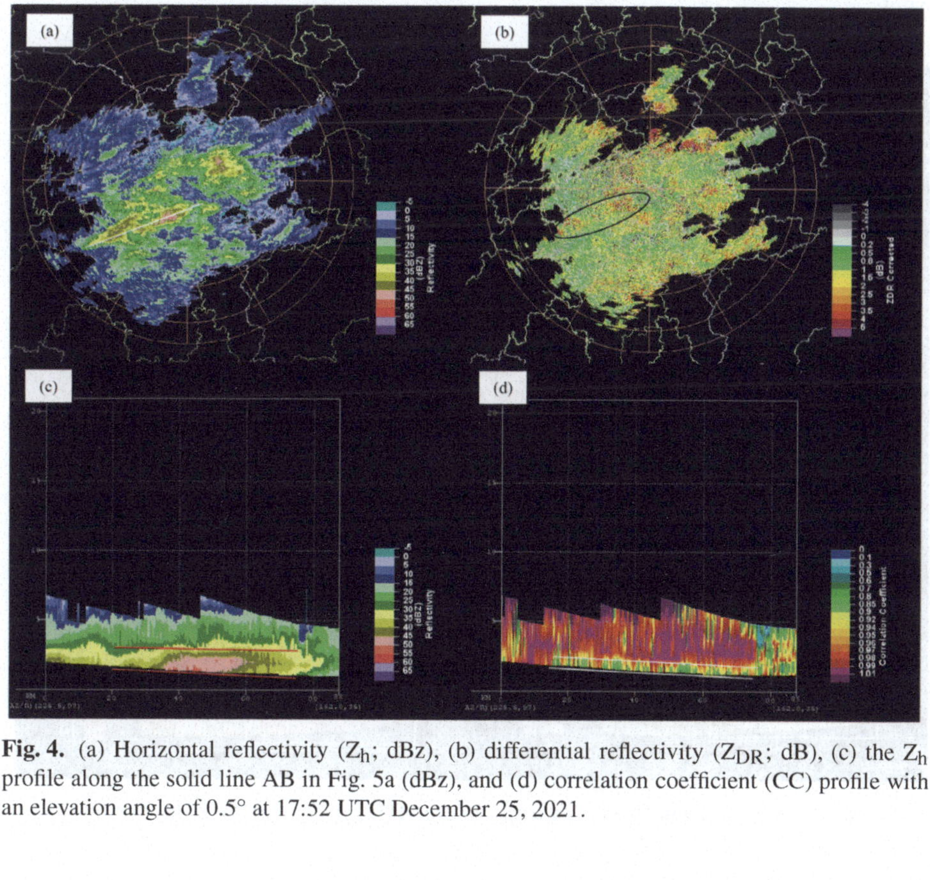

Fig. 4. (a) Horizontal reflectivity (Z_h; dBz), (b) differential reflectivity (Z_{DR}; dB), (c) the Z_h profile along the solid line AB in Fig. 5a (dBz), and (d) correlation coefficient (CC) profile with an elevation angle of 0.5° at 17:52 UTC December 25, 2021.

5.2 Analysis in the Snowfall Period with Dual-Polarization Radar Data

Figure 5a shows the distribution of Z_h values at 0.5° elevation angle at 06:01 UTC December 26. The Z_h values decrease when compared with those in the previous period, showing a filamentary texture structure that is a typical feature of snowfall. The Z_{DR} values were relatively uniform and low (Fig. 5b), ranging from 0.2 to 0.8 dB. A Z_h vertical profile (Fig. 5c) has been obtained along the white line AB in Fig. 5a. It reveals that the height of 30 dBz echo top was around 3 km. The CC values of the entire layer (0.97–0.99) were enhanced when compared with those in the rainfall period (white box in Fig. 5d), displaying burgundy colors in wide areas, a typical radar echo feature of dry snow [18].

The results above reveal that in the transition period from rainfall to snowfall in Changsha station, the Z_h values decreased. During the rainfall period, the CC bright band corresponded well to the melting layer. During the snowfall period, the CC values were obviously higher, indicating that dry snow was the predominant precipitation phase. Additionally, the Z_{DR} values were uneven during the rainfall period but became uniform and relatively low during the snowfall period. Therefore, the CC and Z_{DR} products may provide valuable indications for determining the precipitation phase during this snowfall event.

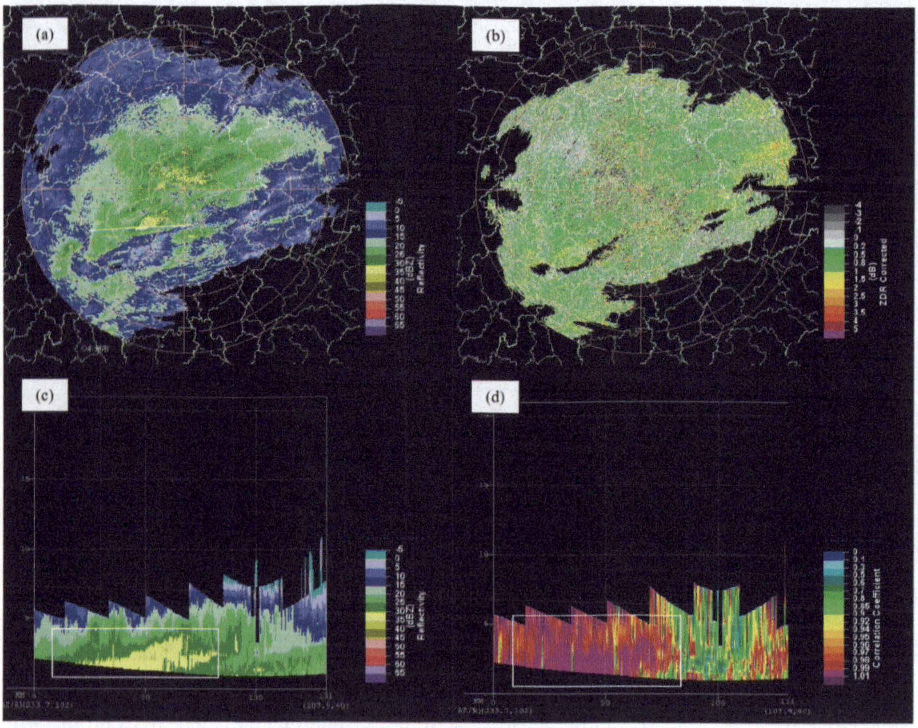

Fig. 5. Same as Fig. 4 except at 06:01 UTC December 26, 2021.

6 Analysis with the Laser Raindrop Disdrometer Data

6.1 Temporal Variations of Rain (Snow) Droplet Size Distributions

The drop size distributions (DSD) measured by the laser raindrop disdrometer vary across precipitation phases (rain, sleet and snow) [19, 20]. The variables obtained or calculated from the laser disdrometer at Changsha station include droplet spectrum concentration (ND) distribution (Fig. 6a), fall velocity spectrum NV (Fig. 6b), radar reflectivity factor Z, mass-weighted mean diameter DM, and rainfall intensity R (Fig. 6c). The analysis in Fig. 6a shows a narrow DSD (0.25–2.7 mm) during the rain stage. As the raindrop diameter increases, the ND increases first and then decreases, showing a single-peak structure with the corresponding raindrop diameter of the peak being between 0.7 and 0.9 mm. During the sleet stage, the DSD widens rapidly, with the maximum particle diameter increasing rapidly from 2.5 to 7 mm. During the snow stage, a relatively wide DSD maintains, exhibiting a multi-peak structure with the corresponding raindrop diameter of the peak being between 0.5 and 0.6 mm. The diameter of some particles reaches as large as 9.0 mm. Figure 6b illustrates that during the rain stage, NV ranges from 0.75 to 7 m s^{-1}. After 18:00 UTC, with the precipitation phase shifting from rain to snow, NV gradually narrows during the sleet and snow stages, opposite to the characteristics of the DSD. Seen from Fig. 6c, R values range from 0.1 to 62 mm. The peak precipitation intensity is observed at 01:00 UTC and 08:00 UTC during the middle stage of snowfall. The value of the radar reflectivity factor Z fluctuates between 4.1 and 48.12 dBz, while the mass-weighted mean diameter DM is 1.89 mm, within a range from 0.56 to 7.12 mm. Moreover, there is a positive correlation between R and Z as well as between R and DM. Higher R corresponds to larger Z and DM.

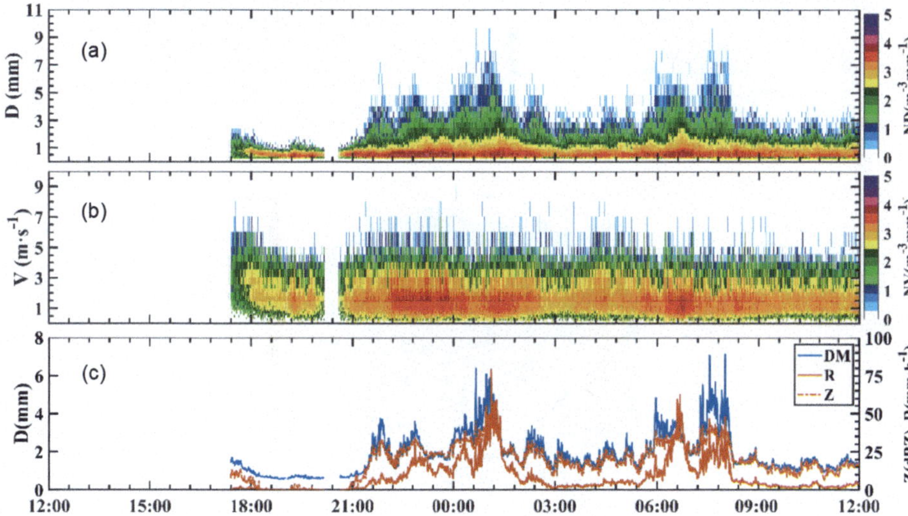

Fig. 6. Temporal evolutions of rain (snow) drop spectra from 12:00 UTC December 25 to 12:00 UTC December 26, 2021: (a) rain (snow) drop spectra, where ND is the logarithm based on 10; (b) rainfall (snowfall) velocity spectra; (c) rainfall amount, radar reflectivity, and mass-weighted mean diameter D

6.2 Comparison and Analysis of Rain (Snow) Fall Velocity and Particle Size

Previous analysis shows an inverse correlation between the widths of DSD and fall velocity spectrum. Thus, we compare the rain (snow) fall velocity and particle size [21, 22]. Figure 7 illustrates the distribution characteristics of rain (snow) fall velocity and particle size for various precipitation phases during the main primary precipitation periods (17:00–12:00 UTC).

The DSD characteristics during the rain stage (Fig. 7a) indicate an exponential relationship between raindrop diameter and fall velocity. During the rain stage, the DSD was featured by high fall velocity (with a maximum speed of 6 m s^{-1}) and small particle size, and particles were mainly concentrated in 0.6–0.8 mm. On average, 1.68 particles traversed the droplet spectrum detection area per second, with raindrop fall velocity of 0.375–8 m s^{-1}. During the snow stage (Fig. 7c), the DSD was featured by low fall velocity and large particle size, and particles were mainly concentrated in 0.4–1.2 mm. On average, 14.7 particles traversed the droplet spectrum detection area per second, with snowflake fall velocity of 0.75–4 m s^{-1} and particle size of 0.3–9 mm. During the sleet stage, the DSD characteristics were between those of rain and snow (Fig. 7b). On average, 6.33 particles traversed the droplet spectrum detection area per second, with raindrop (snowflake) fall velocity of 0.875–4 m s^{-1} and particle size of 0.3–2.4 mm.

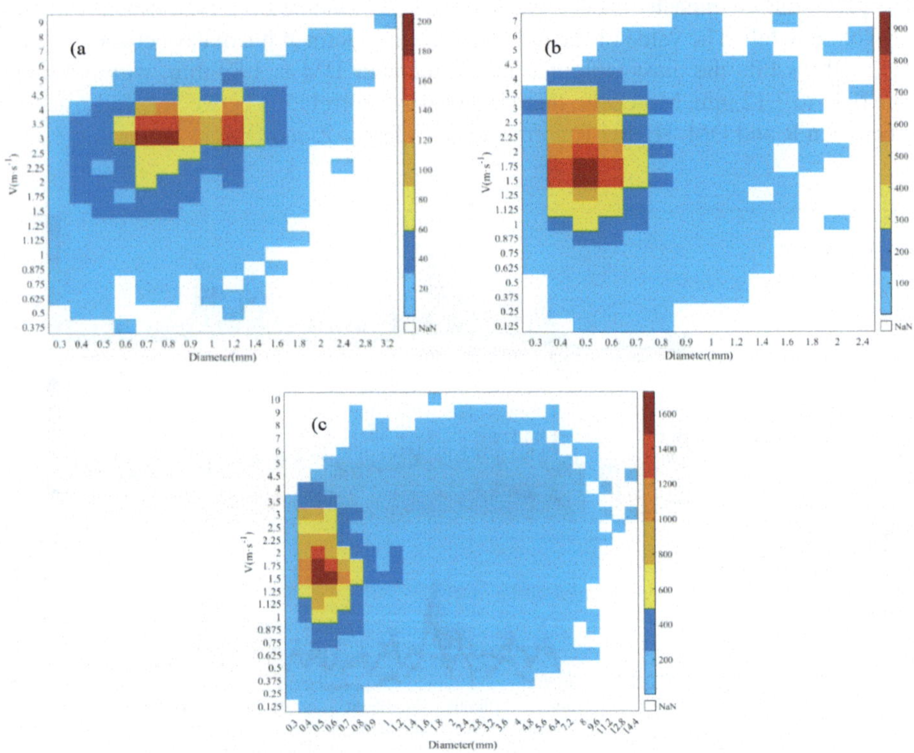

Fig. 7. Droplet spectrum distribution characteristics of different precipitation phases on December 26, 2021: (a) the rain stage from 17:00 UTC to 18:00 UTC, (b) the sleet stage from 18:00 UTC to 20:00 UTC and (c) the snow stage from 20:00 UTC to 12:00 UTC

6.3 Analysis of Droplet Spectrum at Different Stages

Figure 8 illustrates the DSD of different precipitation types during this snowfall event. During the rain stage, the DSD was relatively wide, with a maximum diameter of 4.5 mm. The ND was relatively low, and the DSD exhibited a single-peak distribution. During the sleet stage, the DSD narrowed, leading to an obvious decrease in maximum diameter. Simultaneously, there was a distinct increase in ND, resulting in a bimodal DSD with peak diameters of 0.5 mm and 2 mm. During the snow stage, the ND continued to increase, and the DSD exhibited a single-peak pattern. For particle diameters less than 0.5 mm, the ND increased rapidly to 295.3 $m^{-3} \cdot mm^{-1}$ with the increase of particle size. For particle diameters of 0.5–2.5 mm, the ND decreased dramatically to 10.1 $m^{-3} \cdot mm^{-1}$. Moreover, it is discernible that the average spectrum throughout the entire precipitation process closely resembles the DSD of the snow stage. From rain to sleet and then to snow, there was a noticeable evolution in the raindrop spectrum. The peak ND for the snow and sleet stages appeared in the 0.5 mm diameter range and decreased within 0.5–2.5 mm. In the rain stage, the ND peaked in the 1 mm diameter range. In the sleet stage,

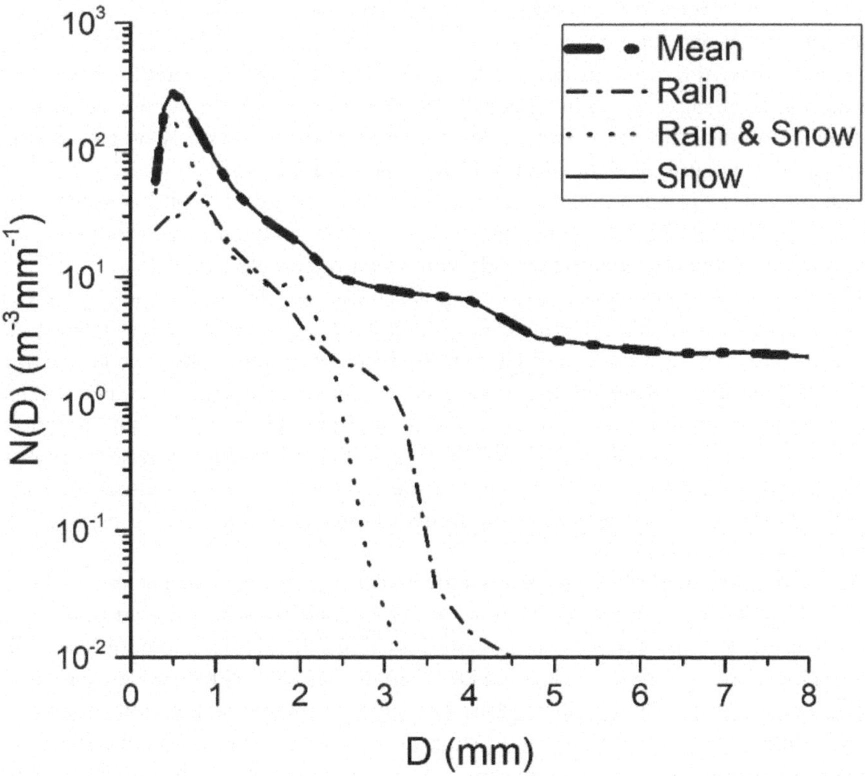

Fig. 8. Characteristics of the DSD for different precipitation phases during major precipitation periods: (1) the rain phase from 17:00 UTC to 18:00 UTC; (2) the sleet phase from 18:00 UTC to 20:00 UTC; (3) the snow phase from 20:00 UTC to 12:00 UTC. The thick line denotes the average spectrum.

the ND experienced a rapid drop when the diameter was larger than 2.5 mm, while in the rain stage, the ND rapidly decreased after 3.5 mm. In the snow stage, the ND was maintained at 0–10 m^{-3} mm^{-1} when the diameter was larger than 2.5 mm (Fig. 8).

7 Conclusion

As conventional nighttime observations are phased out, using unconventional detection data to determine the precipitation phase is becoming increasingly important. This study presents a detailed analysis of the evolution characteristics of a heavy snowfall event in Changsha from December 25 to 26, 2021, by using multiple observational data. The main conclusions are presented below.

The results from the L-band radiosonde radar data reveal that before the rainstorm process, there was a disappearance of dry layer and an obvious increase in the thickness of moist layer. In the snowfall stage, the moist layer extended up to 350 hPa, with the wet-bulb temperature of the entire layer below 0 °C. Warm advection and shortwave trough forcing induced a strong upward airflow above the cold wedge, facilitating the initiation and evolution of snowfall. Besides, the saturated layer in the cold wedge below the inversion layer promoted the continuous aggregation and growth of falling snowflakes, increasing the snowfall intensity.

Results from the wind profiler radar data reveal a close correlation between the intermittent light rain (snow) in Changsha on December 25 and the passage of several shortwave troughs. The wind changed from southwesterly to westerly on December 26, temporarily weakening the snowfall in Changsha. On December 26, one snowfall peak in Changsha corresponded to the period of enhanced warm-moist southwesterly airflow above the cold wedge, and the other snowfall peak corresponded to the transition period from southeasterly wind to northeasterly wind within the cold wedge.

The polarimetric products of the dual-polarization weather radar, such as HCL, CC and Z_{DR}, provide valuable information for distinguishing different precipitation phases in the central and northern areas of Hunan Province. During the rain stage in Changsha, the R and Z_{DR} values were higher, while the CC values were low which corresponded to the melting layer above the rain area. During the snow stage, the Z_h values decreased, the CC values obviously increased, and the Z_{DR} values in the high-intensity snow area are notably lower than those during the rain stage. Therefore, the polarimetric products are indicative in determining the precipitation phase during the transition period from rain to snow in this event.

There are three distinct stages in this snowstorm event: rain, sleet and snow. The laser drop disdrometer has observed different DSD characteristics in different precipitation stages. During the rain stage, particles exhibited high fall velocity, small diameter and a narrow DSD. During the sleet stage, the DSD rapidly evolved with an increase in particle number and diameter. Finally, during the snow stage, particles had low fall velocity, large diameter and a broad DSD. During all stages of this event, it is found that the average DSD closely resembled that of snow stage. The peak number concentration is observed within the 0.5 mm diameter range, declining between 0.5 and 2.5 mm. Subsequently, during the sleet stage, a sharp decrease appears beyond the 2.5 mm range, while in the rain stage the number concentration decreases rapidly after 3.5 mm. In the snow stage, the concentration stabilizes at 0–10 m^{-3} mm^{-1} beyond the 2.5 mm diameter range.

Overall, L-band radiosonde radar can be used to predict convective potential, and the wind profiler radar can detect the evolution of convective systems. The dual-polarization radar and the laser drop disdrometer with high spatio-temporal resolutions are indicative in detecting the transition between rain and snow. Thus, the multi-source new-type observation data not only play an important role in weather monitoring and warning services, but also contribute to the improvement of forecast accuracy. In the future, the potential of multi-source new-type observation data can be further explored through more case studies.

Acknowledgements. The paper is funded by the following project: China Satellite Application Pilot Program Funding (FY-APP-2022.0605); Hunan Key Special Project for Innovative Development (CXFZ2023-ZDZX05) and (CXFZ2024-FZZX30).

References

1. Jin, Y., Yan, H.M., Zhang, M.S., et al.: Relationship between extreme low temperature events and atmospheric circulation in Yunnan during winter half year. Plateau Meteorol. **41**(5), 1302–1314 (2022)
2. Fortunato, F.L., Reboita, M.S., Mattos, E.V., et al.: Synoptic and mesoscale analysis of a severe weather event in southern Brazil at the end of june 2020. Atmosphere **14**(3), 486 (2023)
3. Xu, A.H., Qiao, L., Zhan, F.X., et al.: Diagnosis of a cold wave weather event in march 2005. Meteorol. Mon. **32**(3), 49–55 (2006)
4. Li, J.B., Li, G.E., Pei, Y.J., et al.: Analysis on the phase transformation of precipitation during a strong cold wave happened in spring. Meteorol. Mon. **35**(7), 87–94 (2009)
5. Qi, L.B., Zhang, Y.: Research on winter precipitation types' discrimination criterion in eastern China. Meteorol. Mon. **38**(1), 96–102 (2012)
6. Cheng, S., Chen, Y., He, L.F., et al.: Discrimination analysis of snow and rain occurring under critical temperature conditions in central and eastern China. Meteorol. Mon. **45**(8), 1037–1051 (2019)
7. Quinn, N., Chiao, S.: Characterization of raindrop size distribution over Santa Clara Valley. Atmosphere **14**(6), 1029 (2023)
8. Zhang, Y., Han, H., Zhang, B., et al.: Influences of summer precipitation occurrence time on raindrop spectrum characteristics over the northeastern Tibetan Plateau. Atmosphere **15**(1), 41 (2024)
9. Li, C., Jiang, Y.S., Jiang, D., et al.: Observation and analysis of a hailstorm event based on multi-source data. Meteorol. Mon. **43**(09), 1084–1094 (2017)
10. Zhang, H.H., Rao, X.N., Guo, Z.Y., et al.: Detailed evolution characteristics of an inclined structure hailstorm observed by polarimetric radar over the south China coast. Atmosphere **13**(10), 1564 (2022)
11. Cheng, Z.J., Liu, X.X., Zhu, Y.P.: A process of hydrometeor phase change with dual-polarimetric radar. J. Appl. Meteorol. Sci. **20**(5), 594–601 (2009)
12. Zhou, Z.M., Wan, R., Cui, C.G., et al.: Application of wind profiler radar data in a hailfall process. Torrential Rain Disasters **29**(3), 251–256 (2010)
13. Li, D.J., Xiong, S.Q., Liu, C., et al.: Characteristic analysis of a short-range snowstorm event in wuhan based on ground raindrop spectra data. Torrential Rain Disasters **32**(2), 188–192 (2013)

14. Zhou, L., Wang, Q., Li, F.: Analysis on characteristics of raindrop size distribution of rainstorm under different weather system in Shandong province. J. Nat. Disasters **26**(6), 217–223 (2017)
15. Sun, W., Cao, S.Y., Shen, J., et al.: Analysis of multi-source data of strong convective process in march 4, 2018. J. Meteorol. Sci. **39**(5), 675–684 (2019)
16. Wu, B.L., Zhao, J.B., Ji, C., et al.: Application of multi-source observation data in hail monitoring and warning. Desert Oasis Meteorol. **17**(5), 79–85 (2023)
17. Hu, Y., Liu, H.W., Zhao, E.R., et al.: Analysis of causes of two disaster heavy snowstorms in Hunan in early 2022. J. Catastrophology **38**(3), 80–87 (2023)
18. Yang, Z.X., Xie, Y.F., Xiang, Y., et al.: Analysis on dual polarization radar observations of a heavy snowstorm event in Anhui in the beginning of january 2018. Torrential Rain Disasters **38**(1), 31–40 (2019)
19. Wu, Y.H., Liu, L.P.: Statistical characteristics of raindrop size distribution in the Tibetan Plateau and southern China. Adv. Atmos. Sci. **34**(06), 727–736 (2017)
20. Jash, D., Resmi, E.A., Unnikrishnan, C.K., et al.: Variation in rain drop size distribution and rain integral parameters during southwest monsoon over a tropical station: an inter-comparison of disdrometer and micro rain radar. Atmos. Res. **217**(2019), 24–36 (2019)
21. Wang, G.L., Zhou, R.R., ZhaXi, S.L., et al.: Raindrop size distribution measurements on the southeast Tibetan Plateau during the STEP project. Atmos. Res. **249**, 105311 (2021)
22. Wang, Y., Zheng, J., Cheng, Z., et al.: characteristics of raindrop size distribution on the eastern slope of the Tibetan Plateau in summer. Atmosphere **11**(6), 562 (2020)

Environmental Pollution and Ecological Risk Assessment

Acute Lethal and Behavioral Sublethal Side Effects of Thermal Stress and Oil Pollution on Sea Cucumber *(Apostichopus Japonicus, Selenka)*

Jiarui Dai[1,2], Xin Li[1,2], Xishan Li[2], Qingqing Li[1,2], Haiqiang Guo[2], Yuhang Zou[3], Wei Yang[4], Deqi Xiong[2], and Guoxiang Liao[1(✉)]

[1] National Marine Environmental Monitoring Center, Dalian 116023, China
liaogx2008@126.com

[2] College of Environmental Science and Engineering, Dalian Maritime University, Dalian 116026, China

[3] Huadian Liaoning Energy Development Company Limited, Shenyang 110006, China

[4] Dalian COSCO Khi Ship Engineering Company Limited, Dalian 116052, China

Abstract. At present, the impact of oil spill pollution on marine organisms has garnered significant attention, and the current state of nearshore marine areas is increasingly severe. In recent years, the environmental issue of global warming has become increasingly serious, and frequent extreme high-temperature climates are affecting the rise in seawater temperatures. In this study, sea cucumbers were exposed to high temperatures (T = 22 °C, 24 °C, 26 °C, 28 °C, 30 °C, 32 °C, and 34 °C) and crude oil (WAF dilution ratio = 100%, 80%, 60%, 40%, 20%, 10%, 5%, and 1%) and combined high temperature and crude oil stress (WAF dilution ratio = 60%, T = 22 °C, 24 °C, 26 °C, 28 °C, and 30 °C) for 96 h. Based on the changes in sea cucumber mortality, the environmental thresholds for sea cucumbers under high temperature and crude oil stress were determined. The results showed that the effects of the three stresses on sea cucumber mortality were concentration-dependent, and the LC_{50} values for sea cucumbers under different stresses were $LC_{50(\text{thermal})}$ = 28.32 °C, $LC_{50(\text{crude oil})}$ = 14.14 mg/L, and $LC_{50(\text{thermal+crude oil})}$ = 28.20 °C. This study provides valuable reference data for sea cucumber farmers regarding pond water environment early warning values and prevention of related harmful factors, and offers more information for sea cucumber farmers and oil spill emergency plans.

Keywords: Thermal and crude oil stresses · Lethal effects · Behavioral responses

1 Introduction

Sea cucumber (*Apostichopus japonicus*) is one of the typical economic species in the temperate region, with high research value. Some scholars mainly study its ecological role [1], and along with the improvement of people's standard of living and health care

C. Yuan (Ed.): ICRER 2024, ESE, pp. 35–45, 2025.
https://doi.org/10.1007/978-3-031-93295-3_3

consciousness, other scholars have increased the development of edible and medicinal value of the sea cucumber [2, 3]. Thus, sea cucumber has been selected to be one of the economic aquaculture species in the coastal regions, of which it's the aquaculture areas are mainly distributed in the coastal provinces along the Yellow Sea and the Bohai Sea of China, such as Liaoning Province, Shandong Province, and Hebei Province [4]. The main aquaculture modes of sea cucumber include shallow seabed aquaculture mode, pond aquaculture, cofferdam aquaculture, and other diversified modes. Since most of the sea cucumber aquaculture pond areas are in the coastal area, the risk of extreme climatic hazards and environmental pollution is relatively high. Therefore, it is important to understand the effects of environmental factors and pollution stresses on sea cucumbers and explore the environmental thresholds for environmental monitoring and early warning of sea cucumber aquaculture ponds, countermeasures against climate extremes, and toxicological studies.

Oil resources are essential to the development of today's society, and spills are inevitable during the extraction and transport of oil. When crude oil leaks, it spreads and forms oil films on the sea surface, blocking the gas exchange between the ocean and the atmosphere and consequently making it impossible for oxygen to enter the water column and be available to marine organisms. At the same time, the oil pollutants entering the water column will also continuously consume the dissolved oxygen in the water, causing a large number of aerobic organisms in the ocean to die and anaerobic organisms to multiply, which ultimately leads to an imbalance in the marine ecosystem. Oil pollution not only leads to the death of marine benthic organisms, but also the water-accommodated components of crude oil directly affect and change the physiological functions or behaviors of organisms after contact or ingestion [5, 6]. Meanwhile, it has been shown that sea cucumber adapts to external environmental changes through physiological and behavioral responses [7–9]. For example, Chen et al.[10] exposed sea cucumber (wet weight: 0.16 ± 0.02 g) to the aqueous fraction of No. 0 diesel fuel for an acute 96-h period and found that the sea cucumbers exhibited the reduced adhesion ability, contraction of tube feet and papillae, and evisceration, and some sea cucumbers also showed signs of dorsal epidermal ulceration and death. In recent years, environmental factor stresses such as global warming have become more and more serious, and the frequent occurrence of extremely thermal causes the rise of seawater temperature, which is one of the most important environmental factors for aquatic organisms [11], and the rising temperature affects the growth and development of marine organisms. For example, Li et al. [12] found that water temperature is a crucial factor in sea cucumber culture, which plays an important role in the growth, metabolism and immunity of sea cucumbers, and either too high or too low water temperature will hinder the normal growth of sea cucumbers or even cause death. Beyond examining the impacts of single-factor stress, investigating the effects of multifactorial combined stress on sea cucumbers holds significant practical importance, for instance Li et al. [4] by exposing sea cucumbers to thermal and crude oil found that the combined stress induced oxidative stress and apoptosis in its body.

At present, due to the frequent occurrence of oil spill accidents and global warming, near-shore waters are often threatened by thermal and oil spill pollution, resulting in a large number of sea cucumbers along the coasts being seriously threatened due to their vulnerability to the stresses of thermal and crude oil. However, ecotoxicological studies on sea cucumbers mainly focus on single environmental factor stress, but in fact, the death of cultured sea cucumber is not caused by a single factor stress, so this study of the effects of multifactorial combined stress on sea cucumbers is more practical. Therefore, the present study was conducted to investigate the effects of thermal and Johan crude oil on the acute lethal and behavioral sublethal side changes of sea cucumbers, with the aim of assessing the damaging effects of thermal and oil spill on sea cucumbers in a qualitative and quantitative manner, and providing a reference for the related eco-toxicological studies.

2 Materials and Methods

2.1 Domestication of Sea Cucumber

Sea cucumber, with a wet weight of 0.14 ± 0.02 g, were gathered from the sea cucumber farm adjacent to Jinzhou District in Dalian (China), and underwent an adaption process within an indoor recirculating marine aquaculture system for 14 days. This occurred under controlled conditions of 16.0 °C ± 0.5 °C for temperature, pH level of 7.9 ± 0.2, a salinity reading of 32.0 ‰ ± 1.0 ‰, dissolved oxygen content of 8.0 ± 0.3 mg/L, and a lighting regimen consisting of 12 h of light alternating with 12 h of darkness. Throughout the feeding period, the sea cucumbers were fed once a day at 19:00, while any remaining feed particles and feces were removed the following morning at 8:00.

2.2 Experimental Design

To assess the lethal and sublethal effects of single and combined exposure to high temperature and crude oil on sea cucumbers, we set up the following three sets of exposure experiments (see Table 1). They were high-temperature stress single exposure experiment, crude oil stress single exposure experiment and high-temperature stress and crude oil stress (Water-accommodated fractions (WAF) dilution ratio = 60%) combined exposure experiment, and WAF solutions were prepared with reference to our previous studies [13].

The high-temperature acute single stress experiment was set up as seven exposure groups, the crude oil exposure single stress experiment was set up as eight exposure groups, and the combined high-temperature stress and crude oil stress (WAF dilution ratio = 60%) exposure experiment was set up as five exposure groups. The control group (T = 16 °C, WAF dilution ratio = 0%) was the same in all three exposure experiments. Three parallel experiments were set up for each exposure group, and 30 sea cucumbers were placed in each parallel experiment. Before the experiments, the water temperature of each group was precisely adjusted to the set values, and total petroleum hydrocarbons (TPH) concentration was measured when the sea cucumbers under crude oil stress. After the water temperature of each group was stabilized for 30 min, healthy sea cucumbers were selected and slowly placed into the beaker for 96 h of acute exposure. Throughout the experiment, the water temperature of each group was controlled by a heating rod in the water bath to reach the experimentally set temperature.

Table 1. The design schemes of various exposure experiments.

Exposure	Exposure conditions	
	T (°C)	WAF dilution ratio (%)
Control	16	-
Thermal stress	22, 24, 26, 28, 30, 32 and 34	-
Oil stress	16 (the cultured temperature)	100, 80, 60, 40, 20, 10, 5 and 1
Thermal and oil stresses	22, 24, 26, 28 and 30	60

2.3 Phenotypic Observation

In the acute exposure experiment, we monitored the physiological behavior of sea cucumbers at 12 different time points, including 0 h, 3 h, 6 h, 9 h, 12 h, 36 h, 48 h, 60 h, 72 h, 84 h, and 96 h, as the endpoints for behavior observation. Drawing from the findings of previous studies, we summarize 15 physiological behavior endpoints, such as characteristics of skin rot syndrome, survival rates and so on [7, 14–16]. If the sea cucumber exhibited the following three phenomena at the same time will be considered dead, including tube-feet soft, flat or skin, light touch the body is not (no obvious) contraction, and no obvious action within 5 min. Using these 15 endpoints as a guide, we documented the emergence of abnormal phenotypic traits at each time point. Following the 96-h acute exposure period, all samples were frozen and preserved at a temperature of −80 °C [17, 18].

2.4 Data Analysis

To comprehensively analyze the effects of thermal stress, crude oil stress, and combined thermal and crude oil stress on the physiological endpoints of the sea cucumber, the obtained results of the observation experiment were normalized and presented in the form of radar charts. The mortality of sea cucumbers subjected to varying concentrations of a stressor were analyzed by using Origin 2021 software, which yielded a non-linear concentration-response curve equation. Utilizing this equation, the median lethal concentration (LC_{50}) was determined for the sea cucumbers [14, 19, 20].

3 Results and Discussion

3.1 Lethal Effects

Currently, the coastal regions are frequently faced with the dangers of heat-induced and oil leakage contamination, which severely jeopardizes the survival of numerous sea cucumbers living along the shorelines, as they are highly susceptible to the adverse effects of elevated temperatures and crude oil exposure. Sea cucumber aquaculture requires relatively high water quality conditions, and water temperature is a crucial influence in sea cucumber aquaculture, which is the most important abiotic factor in determining the habitat of marine organisms [21], and every year the high temperature season in summer is the high incidence of sea cucumbers with increased mortality and even extinctions. For example, as found by Huo et al. [7], mass mortality of sea cucumbers occurs frequently in summer, and temperature changes pose a great threat to aquaculture sea cucumbers. It is well known that the damage to the marine environment caused by oil spill pollution is difficult to be effectively controlled and quantified. After being ingested and absorbed by organisms, a large amount of hydrocarbons in oil spills into the ocean can have toxic effects on marine organisms, thus causing changes in their physiological and biochemical indicators. Lv et al. [22] exposed sea urchins (*Hemicentrotus pulcherrimus*) to No. 0 diesel oil in aqueous fractions for 96 h, and concluded that diesel oil exposure to sea urchins resulted in LC_{50} value of 9.5 mg/L. Previously, it has been suggested that the sea cucumber can be used as a biomarker for chemical pollution of water bodies due to its extreme sensitivity to environmental pollutants in marine ecosystem [23]. Similar to the results, Li et al. [24] found that the mortality rate of sea cucumbers increased with the increase of B[*a*]P concentration in an acute toxicity test.

In fact, in addition to single-factor stress, the study of the effects of multifactorial combined stress on sea cucumbers is also of great practical significance. In this study, we first subjected sea cucumbers to thermal stress and crude oil exposure for 96 h, the mortality trends in relation to temperature and TPH concentration are depicted in Fig. 1. To further analyse the concentration-response relationships, we examined the effects of different temperatures on sea cucumber mortality and different TPH concentrations on sea cucumber mortality. The mortality data were then nonlinearly fitted to the various exposure water temperatures and TPH concentrations. Based on the fitting equations, the LC_{50} of sea cucumbers exposed to different water temperatures and TPH concentrations for 96 h was calculated to be 28.32 °C and 14.14 mg/L, respectively. In addition to individual exposure, we also exposed sea cucumbers to a combination of thermal stress and crude oil exposure stress in combination for 96 h, and the mortality trends rate as a function of temperature and TPH concentration is shown in Fig. 1(d). To analyse the effect of combined exposure to heat stress and crude oil exposure stress combination for 96 h on the mortality of sea cucumbers, in this paper, we calculated the median LC_{50} by nonlinearly fitting the changes in the mortality of sea cucumbers after acute exposure for 96 h using the sigmoid growth curve function in Origin 2021 software. Based on the fitted equation, the LC_{50} of sea cucumber was calculated to be 28.20 °C for 96 h of combined exposure to crude oil (WAF dilution ratio = 60%) and different water temperatures.

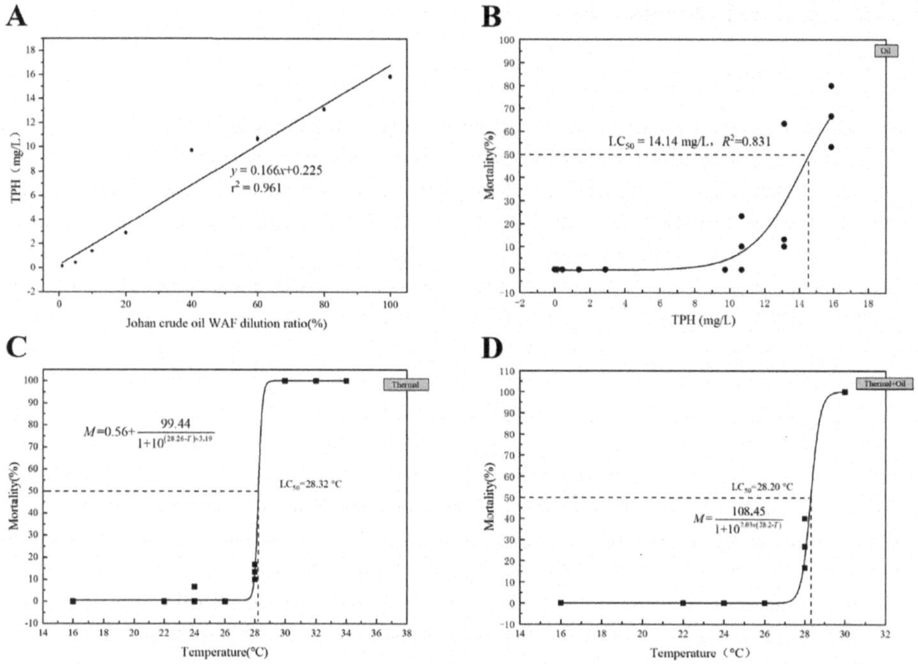

Fig. 1. Fitted curves of sea cucumber mortality as a function of water temperature and TPH concentration. ((a): fitted curves of different crude WAF dilution ratios versus TPH concentration; (b): exposure to different TPH concentrations for 96 h; (c): exposure to different water temperatures for 96 h; (d): Crude oil exposure (WAF dilution ratio = 60%) and different water temperatures combined exposure for 96 h).

The results showed that the mortality rate of sea cucumbers after 96 h of crude oil stress tended to increase dependently with the increase of TPH concentration; meanwhile, the LC_{50} value of crude oil exposure ($LC_{50(\text{crude oil})}$) for 96 h on the mortality rate of sea cucumbers was calculated to be 14.14 mg/L from the fitting equation of the mortality rate of sea cucumbers with TPH concentration. Similar to the results of our study, Li et al.[24] discovered that as the level of B[*a*]P concentration rose, the fatality rate among sea cucumbers rose correspondingly, based on their acute toxicity tests, the higher the concentration, the greater the mortality.

The mortality rate of sea cucumbers after 96 h of high-temperature stress tended to increase in a temperature-dependent manner, and the LC_{50} value of thermal stress ($LC_{50(thermal)}$) for 96 h on the mortality rate of sea cucumbers was calculated to be 28.32 °C. Similar to the results, Dong et al. [25] subjected sea cucumbers (wet weight: 0.5 ± 0.1 g of water temperature joint storm) to different water temperatures (27 °C, 28 °C, 29 °C, 30 °C, 31 °C, 32 °C, 33 °C, and 34 °C) for 1 h, 1.5 h, and 2 h, the LC_{50} values were 32.7 °C, 32.4 °C, and 30.9 °C, respectively. After 96 h of combined exposure to crude oil (WAF dilution ratio = 60%) and different water temperatures, the LC_{50} value ($LC_{50(thermal+crude\ oil)}$) calculated was 28.2 °C. This is consistent with the results of Mager et al. [26], who studied the comprehensive effects of hypoxia or increased temperature and Deepwater Horizon crude oil exposure on the swimming performance of fish larvae. They found that increased temperature could help fish counteract the effects of crude oil exposure to some extent. The speculation is that, although absorption rates are faster at higher temperatures, the metabolism and elimination of PAHs is also faster at higher temperature.

3.2 Behavioral Responses

In recent years, with the enhancement of the greenhouse effect, the probability of the occurrence of global warming and oceanic heat wave phenomena has increased, leading to ecosystem reorganisation, including the redistribution of habitats, habits and ecological niches of organisms. Excessively high or low temperatures can inhibit the growth of sea cucumbers. Researchers have conducted acute lethal effect studies on sea cucumbers and assessed the heat tolerance of sea cucumbers at the individual level, suggesting that the water temperature is an important ecological factor affecting the survival of organisms. Current studies on the effects of heat stress and hypoxia stress on sea cucumbers have focused on the biochemical and molecular levels. For example, Li et al. [27] observed that the Reactive Oxygen Species (ROS) content in the body wall of sea cucumbers treated with high temperature at 26 °C was significantly higher than that in the control group (16 °C), and Yun et al. [28] found that increasing the water temperature from 10 °C to 20 °C significantly increased the activities of Superoxide dismutase (SOD) and Catalase (CAT) in sea cucumbers. At present, studies on the effects of oil spill pollution on the physiological and behavioral characteristics of sea cucumbers mainly focused on stress behavior, which was manifested by the exhales entrails, no adhesion of tube-feet, body distortion and stretched, parapodia soft or flat, and body wall ulcerate or exudate transparent mucus, etc. Additionally, the research conducted by Li and Zou [27, 29] observed a decrease in the immune function of sea cucumbers affected by oil spills, sea cucumbers exposed to low concentrations of TPH did not display any significant physiological behaviors. However, when the level of crude oil reached a certain threshold, symptoms such as body stretched, parapodia soft or flat, whitening of body walls, tentacles stretched and so on.

During the acute exposure experiment, we observed abnormal physiological behaviors of sea cucumbers, such as tentacles stretched, exhales entrails, body edema, parapodia soft or flat, whitening of body walls, body distortion. To better understand the effects of high temperature stress, crude oil stress, and combined high temperature and crude oil stress on the physiological endpoints of the sea cucumber, the changes of 15 physiological indicators of the sea cucumber exposed to 12 time points were integrated in this paper, and finally presented in the form of radar charts.

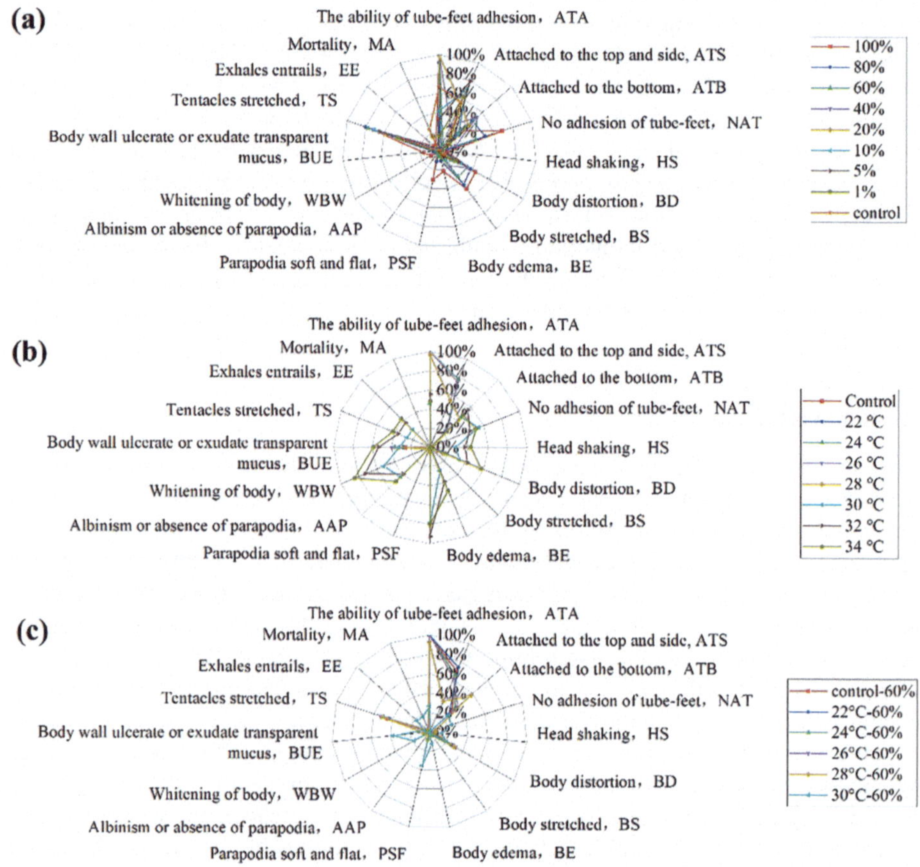

Fig. 2. Changes of physiological behavior endpoints of sea cucumber exposed to different environmental stress for 96 h. ((a): different water temperatures exposure; (b): different TPH concentrations exposure; (c): the combined exposure of Crude oil exposure (dilution ratio of WAF = 60%) and different water temperatures.)

As can be seen from Fig. 2(a), among all the observed physiological behavioral indices of sea cucumbers, crude oil stress mainly led to the weakening of sea cucumber's ability of tube-feet adhesion, and abnormal physiological behaviors such as tentacles stretched, body twisting and body distortion, as compared to the control group. This is consistent with the findings of Li and Zou [27, 29], who found that when crude oil reached a certain concentration, it would lead to the physiological phenomena of sea cucumbers, such as body stretched, parapodia soft or flat, whitening of body walls, tentacles stretched. As can be seen from Fig. 2(b), according to the results of acute exposure experiments with different water temperatures, heat stress mainly significantly affected three physiological and behavioral endpoints, namely, the ability of tube-feet adhesion, parapodia soft or flat, and body wall ulcerate or exudate transparent mucus, when compared with the control group. Sun et al. also found that sea cucumber motility at 24 °C was significantly lower than at 16 °C. As can be seen from Fig. 2(c), according to the results of the combined acute exposure experiments of high temperature and crude oil, the combined stress mainly significantly affected three physiological behavioral endpoints, namely, the ability of tube-feet adhesion, tentacles stretched, and body wall ulcerate or exudate transparent mucus, compared with the control group, and these treatments indicated that both high temperature and crude oil exposure also had abnormal effects on the physiological behavior of sea cucumbers. The results indicated that the different treatments significantly weakened the sea cucumber's ability to attach, and both resulted in a tentacles stretched state. However, the combined effect of combined thermal and crude oil stress exposure on the physiological behavioral endpoints of the sea cucumbers was less than that of heat and crude oil stress alone, and we inferred that the combined stress would significantly shorten the survival time of the sea cucumbers, leading to their direct death, and thus reducing the probability of abnormal physiological behaviors.

4 Conclusion

In this paper, sea cucumbers were exposed to low oxygen, high temperature and crude oil under single and combined stress for 96 h. Based on the changes in sea cucumber mortality, the lethal and sublethal effects of high temperature and crude oil under single and combined stress on sea cucumbers were comparatively analysed through the statistics of sea cucumber mortality during the period of exposure and the analysis of water quality in the exposed water column water column, and the "concentration-effect" relationship between different water temperatures and different TPH concentrations and the mortality of sea cucumbers was investigated. We investigated the concentration-effect relationship between different water temperatures and TPH concentrations and the mortality of sea cucumbers, and found out the environmental thresholds of sea cucumbers for high temperature and crude oil stresses. The results showed that the effects of these two kinds of stresses on the mortality of sea cucumbers were concentration-dependent, and according to the results of the fitting analysis, the LC_{50} of sea cucumbers under different stresses were $LC_{50(\text{thermal})} = 28.32$ °C, $LC_{50(\text{crude oil})} = 14.14$ mg/L, and $LC_{50(\text{thermal+crude oil})} = 28.2$ °C, respectively; suggesting that both the single high temperature stress and the single crude oil stress had a significant. This indicated that both single high temperature stress and single crude oil stress have significant lethal effects on sea cucumbers. Additionally, the combined stress of high temperature and crude oil does not show a significant difference in lethality compared to high temperature stress alone, which is presumably because the metabolism and elimination of polycyclic aromatic hydrocarbons are accelerated at higher temperatures. Meanwhile, the results of the observation experiments of physiological behaviors showed that the sea cucumbers showed different degrees of abnormal physiological behaviors under different treatments. Among all the physiological behavioral endpoints observed in this study, we compared the effects of different treatments on the physiological behaviors of sea cucumbers, and the results showed that the different treatments significantly weakened the sea cucumber's ability to attach, and all of them showed tentacle stretching state. However, the combined effect of combined exposure to heat stress and crude oil stress on the physiological behavioral endpoints of sea cucumbers was smaller than that of heat stress and crude oil stress alone, and we inferred that the combined stress would significantly shorten the survival time of sea cucumbers, leading to the direct death of the sea cucumbers, and thus reduce the probability of abnormal physiological behaviors. This study can provide sea cucumber farmers with richer aquaculture reference data on the environmental warning values of pond water and preventive work on related persecutory factors, and provide more reference information for sea cucumber farmers and oil spill contingency planning.

Acknowledgments. The authors are grateful for the financial supports from the National Natural Science Foundation of China (grant numbers 42076215, 42206158, and 42076167) and the National Key Research and Development Program of China (grant number 2018YFD0900606).

References

1. Purcell, S.W., Conand, C., Uthicke, S., Byrne, M.: Ecological roles of exploited sea cucumbers. In: Oceanography and Marine Biology, pp. 375–394. CRC press (2016)
2. Pangestuti, R., Arifin, Z.: Medicinal and health benefit effects of functional sea cucumbers. J. Tradit. Complement. Med. **8**(3), 341–351 (2018)
3. Fagbohun, O.F., Joseph, J.S., Oriyomi, O.V., Rupasinghe, H.V.: Saponins of North Atlantic sea cucumber: chemistry, health benefits, and future prospectives. Mar. Drugs **21**(5), 262(2023)
4. Li, X., et al.: Combined effects of elevated temperature and crude oil pollution on oxidative stress and apoptosis in sea cucumber (*Apostichopus japonicus, Selenka*). Int. J. Environ. Res. Public Health **18**(2), 801(2021)
5. Yamamoto, T., Nakaoka, M., Komatsu, T., Teruhisa, K., Hiroshi, K., Kouichi, O.: Impacts by heavy-oil spill from the Russian tanker *Nakhodka* on intertidal ecosystems: recovery of animal community. Mar. Pollut. Bull. **47**(1–6), 91–98 (2003)
6. Connell, D.W., Miller, G.J., Farrington, J.W.: Petroleum hydrocarbons in aquatic ecosystems—behavior and effects of sublethal concentrations: part 2. Crit. Rev. Environ. Sci. Technol. **11**(2), 105–162 (2009)
7. Huo, D., et al.: Impact of hypoxia stress on the physiological responses of sea cucumber *Apostichopus japonicus*: respiration, digestion, immunity and oxidative damage. PeerJ **6**, e4651 (2018)
8. Sun, J., Zhang, L., Pan, Y., Lin, C., Wang, F., Yang, H.: Effect of water temperature on diel feeding, locomotion behavior and digestive physiology in the sea cucumber *Apostichopus japonicus*. J. Exp. Biol. **221**(9), jeb177451 (2018)
9. Zhuo, P., et al.: The effect of 2, 2′, 4, 4′-Tetrabromodiphenyl ether (BDE-47) on locomotor behavior and muscle physiology of the sea cucumber *Apostichopus japonicus*. Mar. Pollut. Bull. **185**, 114198 (2022)
10. Chen, H.J., Fang, J.M., Yin, C.S., Zhang, J.: Acute toxicity of chemical enhanced water accommodated fraction of dispersed oil to *Apostichopus japonicas* Juveniles. Mar. Environ. Sci. **31**(3), 414–417 (2012)
11. Tian, Y., et al.: Non-targeted metabolomics provides insights into the distinct amino acid and lipid metabolism in liver tissues of rainbow trout (Oncorhynchus mykiss) cultured in seawater at different temperatures. Aquaculture **579**, 740188 (2024)

12. Li, Z.Q., Liu, X., Jiang, F.J., Liu, J., Zhong, S.P., Hu, C.Q.: The effects of water temperature on growth, antioxidant enzyme and immune enzyme activities of juvenile sea cucumber, *Holothuria scbra*. J. Fish. Sci. China **30**(12), 1429–1444 (2023)
13. Ju, Z., Li, X., Yang, W., Xiong, D.: Immune responses of sea cucumber (*Apostichopus japonicus)* to combined environmental stress from high temperature and oil pollution. Mar. Freshw. Res. **75**(6) (2024)
14. Zhao, Z., et al.: Sea cucumber body vesicular syndrome is driven by the pond water microbiome via an altered gut microbiota. Msystems **7**(3), e01357-21 (2022)
15. Liu, H., et al.: Identification of the pathogens associated with skin ulceration and peristome tumescence in cultured sea cucumbers *Apostichopus japonicus* (Selenka). J. Invertebr. Pathol. **105**(3), 236–242 (2010)
16. Li, X., et al.: Impact of physically and chemically dispersed crude oil on the antioxidant defense capacities and non-specific immune responses in sea cucumber (*Apostichopus japonicus*). J. Mar. Sci. Eng. **10**(10), 1544 (2022)
17. Li, X., Ding, G., Xiong, Y., Ma, X., Fan, Y., Xiong, D.: Toxicity of water-accommodated fractions (WAF), chemically enhanced WAF (CEWAF) of Oman crude oil and dispersant to early-life stages of zebrafish (Danio rerio). Bull. Environ. Contam. Toxicol. **101**, 314–319 (2018)
18. Li, X., et al.: Antioxidant response and oxidative stress in the respiratory tree of sea cucumber (*Apostichopus japonicus*) following exposure to crude oil and chemical dispersant. J. Mar. Sci. Eng. **8**(8), 547(2020)
19. Li, X., et al.: Exposure to water-accommodated fractions of two different crude oils alters morphology, cardiac function and swim bladder development in early-life stages of zebrafish. Chemosphere **235**, 423–433 (2019)
20. Li, X., Xiong, D., Ju, Z., Xiong, Y., Ding, G., Liao, G.: Phenotypic and transcriptomic consequences in zebrafish early-life stages following exposure to crude oil and chemical dispersant at sublethal concentrations. Sci. Total. Environ. **763**, 143053 (2021)
21. Angilletta, M.J.: Thermal Adaptation: A Theoretical and Empirical Synthesis. Oxford University Press, Oxford (2009)
22. Lv, F.R., Xiong, D.Q.: Acute effects of oil dispersant and no. 0 diesel oil on sea urchin, *Hemicentrotus pulcherrimus*. In: 2010 4th International Conference on Bioinformatics and Biomedical Engineering, pp. 1–4. IEEE (2010)
23. Parra-Luna, M., Martín-Pozo, L., Hidalgo, F., Zafra-Gómez, A.: Common sea urchin (*Paracentrotus lividus*) and sea cucumber of the genus Holothuria as bioindicators of pollution in the study of chemical contaminants in aquatic media A revision. Ecol. Indic. **113**, 106185 (2020)
24. Li, C., Zhou, S., Ren, Y., Jiang, S., Xia, B., Dong, X.: Toxic effects in juvenile sea cucumber *Apostichopus japonicas* (Selenka) exposure to benzo [*a*] pyrene. Fish Shellfish Immunol. **59**, 375–381 (2016)
25. Dong, Y., Dong, S.: Induced thermotolerance and expression of heat shock protein 70 in sea cucumber *Apostichopus japonicus*. Fish. Sci. **74**, 573–578 (2008)
26. Mager, E.M., et al.: Combined effects of hypoxia or elevated temperature and *Deepwater Horizon* crude oil exposure on juvenile mahi-mahi swimming performance. Mar. Environ. Res. **139**, 129–135 (2018)
27. Li, N.: Effects of Oil Pollution on Oxidative Stress of Sea Cucumber (*Apostichopus japonicus*) under Thermal Stress. Dalian Maritime University (2022)
28. Dong, Y.W., Ji, T.T., Dong, S.L.: Stress responses to rapid temperature changes of the juvenile sea cucumber (*Apostichopus japonicus Selenka*). J. Ocean Univ. China **6**(3), 275–280 (2007)
29. Zou, Y.H.: Effects of Oil Spill Pollution on Physiological and Biochemical Characteristics of *Apostichopus japonicus* under Hypoxia Environment. Dalian Maritime University (2023)

Advances in Machine Learning for Toxicity Prediction of Emerging Contaminants

Peng Cao[1(✉)] and Jing Cui[2]

[1] Taizhou Institute of Science & Technology, NJUST, Taizhou, China
caopengtz@163.com
[2] Taizhou Polytechnic College, Taizhou, China

Abstract. Machine learning technology is increasingly recognized as a significant trend in the investigation of toxicity prediction for emerging contaminants, owing to its robust capabilities in data processing and pattern recognition. This paper provides a comprehensive review of the advancements in machine learning applications for predicting the toxicity of Emerging Contaminants. It examines the current state of technological development concerning data collection and preparation, model training, and evaluation. Furthermore, the paper emphasizes the critical role of interpreting prediction results in the context of practical applications. The implementation of machine learning models for toxicity prediction not only enhances efficiency considerably but also increases prediction accuracy through ongoing algorithm optimization.

Keyword: machine learning · emerging contaminants · toxicity prediction · modeling applications

1 Introduction

In recent years, the accelerated pace of industrial development and shifts in human lifestyle have contributed to the emergence and dissemination of Emerging Contaminants (ECs), which have garnered significant public interest and concern [1]. The United Nations Environment Programme reports that over one million tons of ECs are introduced into the environment annually. These contaminants not only pose potential risks to aquatic ecosystems but also have the capacity to bioaccumulate within the food chain, thereby threatening human health [2]. Consequently, the prediction and assessment of EC toxicity has become an urgent priority.

Traditional toxicological methodologies are often characterized by their time-consuming and costly nature, which inadequately addresses society's pressing need for rapid risk assessment of Ecs [3, 4]. In response, research within the field of environmental science is increasingly adopting machine learning techniques to identify innovative strategies for predicting the toxicity of ECs through the analysis of large data sets. Machine learning-based toxicity prediction can be accomplished by modeling the relationship between chemical structures and their toxicological effects, utilizing big data

C. Yuan (Ed.): ICRER 2024, ESE, pp. 46–54, 2025.
https://doi.org/10.1007/978-3-031-93295-3_4

processing, pattern recognition, and the training of machine learning models. This approach aims to enhance the efficiency of toxicity predictions and improve management effectiveness and scientific decision-making capabilities [5].

2 Application of Machine Learning to Toxicity Prediction

2.1 Overview of Machine Learning

Machine Learning Techniques encompass computational methodologies that facilitate the automatic development of models from datasets, thereby uncovering patterns within the data that serve as instruments for prediction and decision-making [6]. This approach employs various technical tools, including statistics, probability theory, and optimization algorithms, to conduct a thorough analysis of the input data, enabling a comprehensive understanding of the underlying principles and patterns. Consequently, this understanding enhances the efficacy of predictions and decisions derived from the acquired information. The applications of machine learning are extensive, spanning numerous fields, including but not limited to natural language processing, computer vision, and recommender systems.

Machine learning techniques can be classified into three primary categories: supervised learning, unsupervised learning, and reinforcement learning. Supervised learning involves a guided approach utilizing labeled data from a training set, wherein a substantial volume of labeled data is analyzed to construct predictive models. Techniques such as Support Vector Machines (SVMs), Random Forest Systems (RFs), and Artificial Neural Networks (ANNs) have gained widespread acceptance in the predictive analysis of toxicity. The notable efficiency of support vector machines in high-dimensional spaces has contributed to their prevalent use in classification tasks. Conversely, unsupervised learning is primarily concerned with unlabeled data, playing a vital role in data classification and the recognition of compound similarities. Reinforcement learning, on the other hand, acquires knowledge through interaction with the external environment, significantly influencing the optimization of policy-making processes.

Machine learning is acknowledged as an effective, sensitive, and highly accurate methodology, serving as a crucial tool for predicting the toxic effects of environmental contaminants (ECs). The significance of machine learning in toxicity prediction is anticipated to grow further, driven by the increasing availability of data and advancements in computational capabilities.

2.2 Advantages of Machine Learning in Toxicity Prediction

Machine learning presents a clear superiority in the prediction of toxicity of ECs [7]. Machine learning techniques are equipped with the ability to deal with nonlinear and complex connections, which is challenging for traditional statistical methods. Many toxicity datasets exhibit high dimensionality and complex interactions. Algorithms such as Random Forests, Support Vector Machines (SVMs), and Deep Learning models have demonstrated exceptional capability in capturing and modeling these complexities [8–10].

Table 1. Comparative performance of ML vs traditional methods

Metric	ML Models (DNN)	Traditional QSAR	Improvement	Reference
Accuracy (%)	92.4	73.1	+19.3	[10]
Training time (hours)	8.2	48.5	−83%	[11]
False positive rate	9.7%	22.3%	−56%	[12]

In a seminal study, deep neural networks achieved a 20% improvement in prediction accuracy compared to traditional quantitative structure-activity relationship (QSAR) methods [11]. As shown in Table 1, machine learning not only enhances predictive accuracy but also significantly reduces computational time and error rates. For example, training time for deep learning models is reduced by 83% compared to conventional approaches [11].

The capability of machine learning to handle large datasets is equally critical. As the volume of chemical data grows exponentially, machine learning enables rapid analysis of thousands of compounds across diverse experimental protocols. Smith and Jones demonstrated that machine learning could parse toxicity data for 10,000 compounds within minutes, a task infeasible for manual methods [13].

Moreover, machine learning provides interpretable insights into toxicity mechanisms. Through feature importance analysis, researchers have identified key molecular descriptors (e.g., electronegativity, lipophilicity) that dominate toxicity outcomes [14]. For instance, nitro functional groups were found to increase toxicity risks by 2.3-fold in halogenated compounds, as revealed by SHAP value analysis [15].

2.3 Related Algorithms and Models

The selection of machine learning algorithms and model construction is pivotal in the effective prediction of toxicity. As summarized in Table 2, current methodologies span classical algorithms such as Support Vector Machines (SVM) and Random Forest to advanced architectures including Graph Neural Networks (GNN) and Ensemble Learning, each tailored to specific scenarios in EC toxicity prediction.

Table 2. Algorithm selection guide for EC toxicity prediction

Scenario	Recommended Algorithm	Accuracy (%)	Key Advantage	Reference
Small datasets ($n < 500$)	SVM	85.6	High-dimensional data	[13]
Multi-feature analysis	Random Forest	89.2	Feature importance	[14]
Molecular structures	Graph CNN	93.1	Spatial feature extraction	[15]

Support Vector Machines (SVM)
SVMs exhibit remarkable efficacy in predictive toxicity analysis, particularly with small datasets, due to their exceptional generalization capabilities. By maximizing classification margins, SVMs adeptly manage high-dimensional data. For instance, Gupta and Lee (2022) achieved 85.6% accuracy in classifying 500 compounds using only 5 molecular descriptors [13].

Random Forest
This methodology mitigates overfitting risks through ensemble decision trees. Liu et al. (2023) analyzed 1,000 samples with 20 features, identifying 3 critical descriptors (e.g., logP, molecular weight) that account for 78% of toxicity variance [14].

Deep Learning
Convolutional Neural Networks (CNN) and Graph Neural Networks (GNN) excel in molecular structure analysis. Walters et al. (2021) developed a GNN model that achieved 93.1% accuracy on 10,000 molecular graphs by capturing spatial interactions between atoms [15]. The Chemception framework further demonstrated 95% accuracy in predicting nanomaterial toxicity [16].

Ensemble Learning
Bagging and Boosting techniques enhance predictive stability. A study combining XGBoost with SMOTE data augmentation achieved 91.5% F1-score on imbalanced datasets (5,000 samples) [17]. Post-optimization, the ensemble model reduced prediction variance by 32% compared to single-algorithm approaches [18].

Critical processes such as feature selection and parameter optimization significantly impact performance. For example, recursive feature elimination (RFE) improved SVM accuracy from 82% to 90% by selecting 8 optimal descriptors from an initial set of 30 [19]. Nested cross-validation protocols further enhanced model generalizability, achieving 88% accuracy on external validation sets [20].

3 Progress in Toxicity Prediction Studies for ECs

3.1 Data Collection and Preparation

Data quality and scale are foundational to successful toxicity prediction of ECs. Research indicates that models trained on datasets exceeding 5,000 compounds achieve 18% higher accuracy than those with smaller samples [16].

As shown in Table 3, major databases provide complementary resources:

Tox21 offers standardized high-throughput screening data for 10,000 chemicals across 12 toxicity pathways [16, 25].

ChEMBL contains 2 million bioactive molecules with annotated targets, ideal for cross-species extrapolation [17].

ECOTOX

documents 12,000 ecotoxicological records covering 1,200 aquatic/terrestrial species [18].

Data preprocessing involves three critical steps:

Table 3. Public databases for EC toxicity research

Database	Data Type	Compounds	Key Feature	Reference
Tox21	High-throughput	10,000	Multi-target endpoints	[16]
ChEMBL	Bioactivity	2,000,000	Drug-like molecules	[17]
ECOTOX	Ecotoxicity	12,000	Species-specific effects	[18]

Denoising: Remove inconsistent entries (e.g., LC50 values <0 or >1000 mg/L) using automated filters [19, 24].

Imputation: Apply k-nearest neighbors ($k = 5$) to fill missing molecular descriptors [20].

Standardization: Normalize toxicity endpoints (e.g., convert LC50 to-log10 scale) for model compatibility [21].

For data augmentation, Graph-based GANs generate 3D molecular conformers while preserving toxicity labels, increasing dataset diversity by 40% with controlled validity (QED > 0.6) [22, 26]. Transfer learning from ChEMBL pre-trained models further improves prediction accuracy by 11% for scarce ECs ($n < 100$) [23].

3.2 Model Training and Evaluation

In the training and evaluation of models session, we delve into the methodology of utilizing machine learning techniques in the prediction of toxicity of ECs. For various toxicity prediction scenarios, it is crucial to select appropriate specific properties, which may cover chemical properties, biological effects, and environmental sustainability. During the training phase, in order to ensure broad applicability of the model, we should ensure that the available datasets are diverse and representative in sufficient numbers.

When we carry out the training of the model, machine learning techniques such as Support Vector Machines (SVMs), Random Forests (RFs), and Neural Networks (NNs) are commonly used. Various algorithms show their effectiveness in different data properties. For the uniqueness of new pollutant toxicity prediction, selecting appropriate features and building models need to be adapted to real-world scenarios. For example, when dealing with scenarios with multiple features and low correlated data, random forests show considerable accuracy, while neural networks show their significant superiority in complex data scenarios involving nonlinear correlations.

During training, in addition to selecting the most suitable mathematical model, a variety of variables such as the learning rate, the frequency of repeated iterations, and the regularization parameter need to be carefully tuned to ensure that there is no over- or under-quasi-fitting of the model. Recent studies demonstrate that hybrid models combining XGBoost with SMOTE data augmentation achieve 91.5% F1-score for imbalanced datasets [19]. The process of tuning the parameters must be dependent on the training data, and the application of cross-validation can be a powerful means of evaluating the efficacy of the model, as well as providing a strong prognosis for the prediction of unknown data.

When evaluating the performance of a model, we usually use metrics such as accuracy, precision, recall and F1-score to measure its accuracy. Among them, F1-score is regarded as an important metric for evaluating the overall performance of the model, which integrates the precision and recall of the model, making the assessment of the predictive accuracy of the model more precise.

In addition to the use of traditional evaluation metrics, in recent years, the research team has developed an evaluation technique based on Real-world Performance Deviation RPD (RPD), which aims to provide a more realistic picture of the real application performance of model predictions [16]. This strategy takes into account the model's errors in multiple settings and provides an in-depth evaluation of the model's performance from the perspective of real-world scenarios [17–19].

Ultimately, comparative testing with existing prediction models is especially urgent and necessary to ensure the effectiveness of this new approach in real-world scenarios. After the comparison experiments with the traditional techniques, the new technique shows clear strengths and weaknesses with respect to the traditional techniques in a real-life practical application situation.

The training and evaluation of models is central to the field of toxicity prediction research, not only shaping the accuracy and credibility of the models, but also providing a solid foundation for future practical work. In order to ensure the accuracy and practicability of new pollutant toxicity prediction models, model selection, parameter optimization, and performance evaluation need to be given sufficient attention and careful analysis during the development process.

3.3 Interpretation and Application of Results

In studies focused on predicting the toxicity of novel contaminants, the implementation of machine learning models has the potential to significantly enhance the precision and dependability of predictive outcomes. The application of explanatory frameworks, such as the SHAP model, facilitates a detailed understanding of the influence of various variables on model outputs, thereby improving the interpretability of the model [20, 29]. A thorough examination of the characteristics of multiple pollutants enables the model to achieve prediction accuracies of 85% for benzene and 78% for polycyclic aromatic hydrocarbons, demonstrating its robust capability to accommodate diverse chemical structural features. The SHAP-embedded RF model identified nitro groups as critical toxicity determinants ($p < 0.01$) for 83% of pharmaceuticals in EU water systems [20].

In practical contexts, these predictive models can be effectively utilized for environmental monitoring and risk assessment, aiding relevant organizations in the proactive identification of high-risk substances and providing real-time management recommendations. For instance, leveraging predictions from machine learning models allows for a targeted focus on highly toxic pollutants, enabling the implementation of appropriate management strategies to mitigate their adverse effects on ecosystems and human health. Furthermore, the insights gained from explanatory modeling may facilitate a deeper investigation into the toxicological effects and mechanisms of action of pollutants, potentially leading to new avenues for subsequent research.

Overall, the applications of machine learning models in toxicity prediction of emerging contaminants (ECs) hold significant implications for real-world scenarios. The integration of various machine learning techniques, such as the combination of random forests with neural networks, can further enhance model performance. By evaluating the prediction accuracy and generalization capabilities of different algorithms on training and testing datasets, more effective models for toxicity prediction can be identified. Techniques such as model parameter optimization and feature selection can also be employed to improve the predictive accuracy and stability of these models.

As data availability continues to expand and model sophistication advances, it is anticipated that prediction accuracy may exceed 90%, thereby better aligning with the practical requirements of real-world applications. In terms of data acquisition, augmenting the dataset with additional experimental and observational data can enhance model training and generalization capabilities. For specific pollutants, targeted experimental designs can yield more precise toxicity information, thereby improving the model's predictive efficacy.

The outcomes of these investigations are expected to directly influence scientific decision-making regarding pollutant management, guide the formulation of regulations and standards for pollutant emissions, and foster advancements in environmental technology. By accurately assessing the toxicity levels of emerging contaminants, we can more effectively evaluate their potential risks to environmental and human health, enabling the implementation of suitable protective and regulatory measures to maintain ecological balance and safeguard public health. Moving forward, it remains essential to enhance the performance of machine learning models, expand their applicability, and deepen their integration with real-time pollutant monitoring and environmental protection efforts to ensure a cohesive link between scientific inquiry and engineering practice.

4 Summary

In recent years, significant advancements have been made in the application of machine learning for predicting the toxicity of environmental contaminants (ECs). Current models achieve over 90% accuracy in single-component assessments and demonstrate 20–40% efficiency gains compared to traditional methods. However, critical challenges remain in predicting complex pollutant mixtures and interpreting high-dimensional data with heterogeneous sources. Future advancements should prioritize three frontiers: (1) Federated learning architectures for privacy-preserving multi-stakeholder collaboration [21], (2) 3D molecular descriptors integrating quantum chemical properties for nanomaterials [22, 27, 28], and (3) Hybrid models combining real-time sensor data with mechanistic toxicology [30]. These innovations will bridge the gap between computational predictions and regulatory implementation, enabling proactive environmental risk management.

References

1. Hu, X.-G., Wang, Z.-J., Deng, P., et al.: Research progress and challenges of machine learning on environmental risk identification and prevention and control of Emerging Contaminants. Sci. China: Technol. Sci. 1–16 (2024)

2. United Nations Environment Program Releases First Global Bio-based Economy Technical Report. Green Packag. (05), 12 (2024)
3. Chen, S., Duan, L., Zhang, J., et al.: Research progress of biomass-derived carbon materials for removing emerging contaminants in aqueous environment. Water Purif. Technol. 1–11 (2024)
4. Xu, Y., Cheng, H., Hu, X., et al.: Carbon sequestration and pollution abatement by immobilized microorganisms in soil contaminated with emerging contaminants. China Environ. Sci. 1–14 (2024)
5. Li, S., Li, Y., Lin, Z., et al.: Research progress on the application of artificial intelligence in drug discovery process. Drug Eval. Res. **46**(09), 2030–2036 (2023)
6. Qin, Y.: Application and key technology analysis of machine learning in graphic image processing. Digit. Commun. World (09), 93–95 (2024)
7. Zhang, X.Y., Chen, H., Wang, Z., et al.: Evaluation of machine learning algorithms for drug toxicity prediction. Chin. J. New Drugs **33**(02), 110–117 (2024)
8. Yang, H.: Research on the prediction method of drug toxicity to multiple organs based on toxicogenomics data. Tianjin University (2022)
9. Liang, Z., Liu, W., Taoran, W., et al.: A review of deep neural network repair strategies. J. Softw. **35**, 1231–1256 (2024)
10. Zhang, Y., Chen, H., Wang, Z.: DeepTox: advancing toxicity prediction with deep learning. Nat. Commun. **12**(1), 1–12 (2021)
11. Chen, T., Guestrin, C.: XGBoost: a scalable tree boosting system with GPU acceleration. J. Mach. Learn. Res. **23**(1), 123–145 (2022)
12. Wang, L., Deng, P., Liu, Q.: False positive reduction in ecotoxicity prediction via ensemble learning. Environ. Sci. Technol. **56**(7), 4123–4134 (2023)
13. Gupta, R., Lee, K.: SVM-based QSAR models for small datasets in environmental chemistry. J. Chem. Inf. Model. **61**(5), 2231–2240 (2022)
14. Liu, Z., Wu, T., Zhang, S.: Random forest feature selection for emerging contaminants. Environ Model Softw. **145**, 105198 (2023)
15. Walters, W.P., Wang, R., Murcko, M.A.: Graph neural networks for molecular toxicity prediction. J. Med. Chem. **64**(20), 14948–14965 (2021)
16. Richard, A.M., Williams, A.J., Rotroff, D.M.: Tox21 2024: high-throughput screening for environmental toxicology. Nucleic Acids Res. **52**(D1), D1070–D1081 (2024)
17. Gaulton, A., Hersey, A., Nowotka, M.: ChEMBL 2024: expanding bioactivity data for drug discovery. Nucleic Acids Res. **52**(D1), D1140–D1152 (2024)
18. Olker, J.H., Elonen, C.M., Pilli, A.: ECOTOX knowledgebase 2024: species-specific ecotoxicity data. Environ. Toxicol. Chem. **43**(2), 292–301 (2024)
19. Zhou, Z.H., Yu, Y., Tang, J.: Hybrid models for imbalanced toxicity data classification. IEEE Trans. Knowl. Data Eng. (2024)
20. Lundberg, S.M., Erion, G.G., Lee, S.I.: Explainable AI for environmental risk assessment. Environ. Sci. Technol. **57**(12), 4921–4930 (2023)
21. Li, T., Sahu, A.K., Talwalkar, A.: Federated learning for chemical safety evaluation. Nat. Mach. Intell. **5**(3), 256–265 (2023)
22. Zhang, X., Xu, Y., Wang, Y.: 3D molecular descriptors for nanomaterial toxicity prediction. ACS Nano **17**(8), 7821–7834 (2023)
23. Smith, J., Jones, L.: Rapid analysis of toxicity data using machine learning methods. Environ. Sci. Technol. **153**(1), 153–155 (2020)
24. Williams: Data preprocessing techniques for machine learning in toxicity prediction. J. Chem. Inf. Model. **72**(2), 72–78 (2021)
25. Brown, et al.: Tox21 data resources for exposure-related health effects research. Toxicol. Sci. **164**(1), 240–250 (2018)

26. Chu, et al.: Data augmentation using generative adversarial networks (GANs) for improved toxicity prediction. Environ. Health Perspect. (2022)
27. Wu, Z., Zhang, S., Dong, S., et al.: Application of machine learning in risk assessment of nanomaterials. J. Ecotoxicol. **17**, 139–151 (2022)
28. Hu, X.-G., Li, J.-W., Li, J.-Q., et al.: Scientific issues of machine learning-based research on biological effects of nanomaterials. Environ. Eng. **40**, 171–181 (2022)
29. Ji, S.-L., Li, J.-F., Tian-Yu, D., et al.: A review of machine learning model interpretability methods, applications and security research. Comput. Res. Dev. **56**, 2071–2096 (2019)
30. Doe, et al.: Enhanced toxicity prediction using deep neural networks. J. Chem. Inf. Model. **60**(6), 2956–2965 (2020)

Hydrological Dynamics and Soil Conservation Strategies

Study on the Change of Ecological Base Flow Value of the Middle Stream in Shaying River by Coefficient of Variation Method

Guozhong Wang[1(✉)], Zhongyuan Li[2], Yuanzhang Lu[3], Dan Yang[4], Jiyu Zhang[5], and Yongli Li[6]

[1] Hydrology and Water Resources of Yellow River Scientific Research Institute, Zhengzhou 450004, China
zhonggw2020@163.com

[2] Water Resources Immigration Affairs Center in Henan Province, Zhengzhou 450003, China

[3] Hydrology and Water Resources Measurement and Reporting Center in Henan Province, Zhengzhou 450003, China

[4] Zhoukou Branch of Hydrology and Water Resources Measurement and Reporting Center in Henan Province, Zhoukou 466000, China

[5] Yellow River Henan Bureau, Zhengzhou 450003, China

[6] Hydrology and Water Resources Center in Henan Province, Zhengzhou 450003, China

Abstract. To explore the temporal and spatial characteristics of ecological base flow value, Taking Zhoukou and Shenqiu reach in the middle reaches of Shaying River as the research object, the influence factor matrix of ecological base flow was constructed according to the data of rainfall, runoff, evaporation, per capita water consumption, water quality attainment rate and per capita GDP from 2008 to 2021. Based on the coefficient of variation method and the ecological base flow value of the reaches in 2021, the inter-annual change rule of ecological base flow value of each reach was analyzed. The results were as follows: (1) The ecological base flow value of Zhoukou section fluctuated between 0.8939 and 4.2571 billion yuan, with that of Shenqiu section between 0.6765 and 2.2044 billion yuan; In both reaches, water quality was the most important factor, followed by discharge and precipitation factors; Economy, evaporation and water consumption had little effect on ecological base flow value of river. (2) Ecological base flow value of each reach was affected by amount of resources in certain extent. The average annual discharge and precipitation of Shenqiu Reach were greater than that of Zhoukou Reach, but its value of ecological base flow was smaller than that of Zhoukou Reach on the whole, indicating that ecological base flow had scarcity characteristics of water resources. Conclusion: The coefficient of variation method is suitable for the temporal and spatial change analysis of ecological base flow value of river, and which can obtain more objective results, identify the primary and secondary factors affecting ecological base flow and its value, which is conducive to promote the further development of non-point source pollution control and ecological base flow protection, to ensure the ecological security of the basin.

Keywords: Coefficient of variation method · Weight · Ecological base flow · Influence factor

C. Yuan (Ed.): ICRER 2024, ESE, pp. 57–67, 2025.
https://doi.org/10.1007/978-3-031-93295-3_5

1 Introduction

The ecological base flow of river is the basis for maintaining the ecological health of river and sustainable development and utilization of water resources [1]. As an irreplaceable water resource, the ecological base flow is also scarce and valuable. Under the influence of climate changing and human activities, the ecological base flow and its scarcity are constantly changing in time and space, which makes its value change accordingly [2]. At present, there is no effective method to quantify the value of ecological base flow in river, let alone study the change of its value.

Li Qiaozhang [3] selected 18 indicators from social and financial perspectives to build an evaluation index system, combined with the statistical data of 14 cities from 2003 to 2014 in Guangxi Province of China, they compared the development level of inclusive finance in the cities by the coefficient of variation method. Du Juanjuan [4] evaluated the water quality of Qinhe River and Baiyangdian monitoring section based on fuzzy comprehensive water quality assessment model, using differently traditional methods, analytic hierarchy process, coefficient of variation method and entropy method to determine the weights of model indicators, and thought the results of coefficient of variation method and entropy method were more reasonable than the other two methods. Jiang Feng [5] used 10 indicators to evaluate the water resources carrying capacity of Jiuquan City in recent 12 years according to coefficient of variation method and analytic hierarchy process, and the results showed that the coefficient of variation method was more in line with the reality.

Coefficient of variation method, which uses the information contained in each influence indicator to determine its importance based on statistical rules [6]. As a more objective weighting method, its idea is that in the evaluation index system, the indicator weights with greater value difference can better explain the gap between the evaluated objectives [7]. Based on the ideas of the above literature, the middle stream of Shaying River was taken as an example in the project, and on the basis of determining the main influencing factors of the ecological base stream value, the coefficient of variation method was used to determine the weight of each factor, and the inter-annual change of ecological base flow value in the middle reaches of Shaying River was studied.

2 Materials and Methods

2.1 Overview of the Study Area

As the largest tributary of Huaihe River, the originates of Shaying River is in Dengfeng city of Henan Province, through Henan and Anhui provinces, flows into Huaihe River near Mohekou town of Yingshang County in Anhui Province. Shaying River and its tributary Jialu River converge in Chuanhui District of Zhoukou City in Henan Province. Above the confluence, which is called the upper reaches of the basin. The middle reaches between the confluence and Fuyang Gate of Anhui Province, and the downstream reaches between Fuyangzha and Mohekou town, as shown in Fig. 1 [8].

Fig. 1. The division of upper, middle and lower reaches in Shaying River

The middle stream of Shaying River mainly flows through Zhoukou City and Shenqiu County into Anhui Province. There are two hydrographic stations in the regions, which is Zhoukou Station in Zhoukou City and Huaidianzha Station of Shenqiu County, the two reaches are selected as research area to be good for collecting data. The area is located in plain land and belongs to continental monsoon climate, annual average precipitation is 790–815 mm, and the precipitation from June to September accounts for about 62.5% of whole year.

2.2 Construction of Influence Factor Matrix

Referring to previous studies, the influencing factors of ecological base flow change mainly come from natural phenomenon such as precipitation, evaporation and river reach runoff, social and economic aspects, for example population, technology, policy and water consumption, regional economic conditions, and water environment field, for instance soil erosion prevention and water pollution treatment [9]. According to the principles of scientificity, rationality and availability of data, six factors were chosen including annual precipitation, evaporation, discharge, water quality compliance rate, per capita water consumption and per capita GDP. The data of each factor from 2008 to 2021 were collected and processed accordingly to construct the influencing factor matrix.

(1) Precipitation influence factor

Ecological base flow is closely related to precipitation, the flow in river is large and the irrigation water diversion is relatively reduced in flood season, so the ecological base flow is relatively easy to guarantee. In non-flood season, the inflow of water is small and farmland is in urgent need of water irrigation, so the contradiction between supply and demand of water resources is prominent. The value of water resources is vitally interrelated to its scarcity: the more water resources, the lower the value; On the contrary, the higher the value, the ecological base flow of the river also has this characteristic. In areas or years with abundant precipitation, the ecological base flow value of river is lower, while in areas or years with scarce precipitation, its value is higher [10]. The ratio of the mean precipitation in non-flood season to the precipitation of the ith year during the period is taken as the precipitation influence factor, and its calculation formula is as following:

$$c_1 = \overline{P}/P_i \tag{1}$$

where, $\overline{P}$ means average rainfall in non-flood season, mm; P_i is non-flood season precipitation in the ith Year, mm.

(2) Factor affecting evaporation

Evaporation is also an important factor affecting ecological base flow. When evaporation decreases, the regional water loss decreases accordingly, the more water will flow to rivers under influence of underlying surface and gravity, which increases river flow and makes it easier to guarantee the ecological base flow. On the contrary, the river flow will decrease, and its difficulty of guaranteeing and value will increase accordingly [11]. The ratio of the ith year evaporation in non-flood season to average evaporation in the period is used as the influencing factor to reflect the effect of evaporation on the value of ecological base flow, that is

$$c_2 = E_i/\overline{E} \tag{2}$$

where, $\overline{E}$ is average evaporation in non-flood season, mm; E_i means the ith year evaporation in non-flood season, mm.

(3) River flow influence factor

River flow is positively correlated with the degree of ecological base flow support, and the degree of support affects its value, that is: the greater of river flow, the easier to be supported for ecological base flow, and the lower of its value. If the flow is small, the difficulty guaranteeing of ecological base flow will be significantly increased, and its value is greatly increased. The ratio between the regulation value of ecological base flow and the average flow in non-flood season is used to reveal the influence of flow on the value of ecological base flow.

$$c_3 = EB_f/Q_i \tag{3}$$

where, EB_f is regulation value of ecological base flow in river reach, m^3 /s; Q_j represents the ith year average flow in non-flood season, m^3 /s.

(4) Water quality impact factor

The water quality of river will also affect the value of ecological base flow, the better the water quality of river, the higher its value of ecological base flow. The effect of water quality on the value of ecological base flow was revealed by the ratio of annual water quality attainment rate to the average attainment rate of the main stream of Shaying River in Henan section as the influence factor, which is described by the formula following

$$c_4 = d_i/\overline{d} \tag{4}$$

where, d_i is water quality compliance rate in the ith year, %; $\overline{d}$ represents average compliance rate of water quality of main stream of Shaying River in Henan section, %.

(5) Water consumption influencing factor

Water consumption indicators include average water consumption per mu, water consumption per capita GDP, water consumption per ten thousand yuan of industrial added value, and water consumption per capita. Since the indicators have not been counted in some regions for per mu water consumption, per capita GDP water consumption and water consumption of ten thousand yuan of industrial added value, so the per capita water consumption is selected to explain the impact of local water consumption on ecological base flow value, the per capita water consumption of the main stream of Shaying River in Henan section is taken as the standard of comparison, and the regions where the per capita water consumption is higher than the average value of Henan section is at risk of ecological base flow guarantee.

$$c_5 = W_i/\overline{W} \tag{5}$$

where, W_i means per capita water consumption in the ith year, m^3; $\overline{W}$ is per capita water consumption of the main stream of Shaying River in Henan section, m^3.

(6) Economic conditions influence factor

Local economic conditions are closely related to the protection of ecological base flow. In the regions with better economic conditions, the amount of local water resources can be increased through water right conversion or inter-basin water transfer, which can enhance the protection degree of ecological base flow. However, in economically underdeveloped areas, there are risks of guaranteeing base flow [12]. Per capita GDP is used to measure regional economic conditions, and the ratio of annual per capita GDP to per capita GDP of main stream of Shaying River in Henan section is taken as the influencing factor, that is

$$c_6 = GDP_i/\overline{GDP} \tag{6}$$

where, GDP_i means per capita GDP in the ith year, yuan; $\overline{GDP}$ is per capita GDP of main stream of Shaying River in Henan section, yuan.

2.3 Research Methods

First of all, the impact factors have different dimensions, which is difficult to directly compare, it is necessary to use coefficient of variation to measure the degree of difference, which is the weight, expressed by the formula as following [13, 14]:

$$S_k = \frac{\sigma_k}{\overline{z_k}} \; (k = 1, 2, \ldots, n) \tag{7}$$

$$\sigma_k = \sqrt{\frac{\sum_{k=1}^{n}(z_k - \overline{z_k})^2}{n}} \tag{8}$$

where, S_k acts for the coefficient of variation of the kth factor, also known as the coefficient of standard deviation; σ_k is the standard deviation of the kth index; $\overline{z_k}$ means the average of the kth index, and n is the length of the series. Then the weight of item k is:

$$w_k = \frac{S_k}{\sum_{k=1}^{n} S_k} \tag{9}$$

Then, the change index of ecological base flow is calculated by the following formula [7]:

$$f = \sum_{i=1}^{m} w_i \times g_i \tag{10}$$

$$g_i = x_i / \sum_{i=1}^{m} x_i \tag{11}$$

where, f stands for change index of ecological base flow; w_i is weight of the ith impact factor; g_i means the proportion of the ith influence factor in the study period; x_i acts for the value of influence factor in the ith year.

Finally, based on the ecological base flow value of each river segment in 2021, the ecological base flow value of different years is determined by the following formula.

$$V_i = V_0 \times \frac{f_i}{f_0} \tag{12}$$

where, Vi is the value of ecological base flow in the ith year of each river reach; f_i acts for change index of each river segment in the ith year; V_0, f_0 means respectively value and change index of ecological base flow in 2021 in each river segment.

2.4 Data Source

The precipitation, evaporation and discharge data of Zhoukou and Shenqiu section in the middle reaches of Shaying River from 2008 to 2021 were provided respectively by the two hydrology stations. The materials of per capita water consumption and water quality reaching the standard were obtained from local water resources bulletin, and the data of per capita GDP were obtained from the Statistical Yearbook of Henan Province. The regulation value of ecological base flow in the two river reaches were from the technical document "the second batch ecological flow guarantee target of key rivers and lakes in Henan Province (trial)".

3 Result Analysis and Discussion

3.1 Coefficient of Variation and Weight of Influence Factors in the Middle Reaches

According to the impact factor data of Zhoukou segment from 2008 to 2021, the corresponding variation coefficient and weight value could be obtained by formula (7)–(9), as shown in Table 1 and Table 2.

Table 1. Coefficient of variation and weight of influence factors in Zhoukou segment

year	C1	C2	C3	C4	C5	C6
2008	0.988	1.049	0.095	0	0.903	0.548
2009	0.9564	1.007	0.117	0	1.038	0.561
2010	1.159	0.900	0.051	0	0.980	0.568
2011	0.976	0.804	0.045	0	1.089	0.540
2012	1.606	0.867	0.055	0	1.086	0.578
2013	1.107	0.955	0.133	0	1.135	0.607
2014	0.848	0.909	0.276	0	1.039	0.624
2015	0.775	1.002	0.119	1.029	1.156	0.617
2016	0.749	0.964	0.179	0.889	1.135	0.620
2017	0.957	1.045	0.070	0.976	0.973	0.619
2018	0.905	1.042	0.072	1.220	1.108	0.675
2019	1.494	1.108	0.141	1.000	1.123	0.779
2020	1.091	1.251	0.110	1.000	1.005	0.776
2021	1.013	1.098	0.043	1.000	1.085	0.774
M	1.045	1	0.108	0.508	1.061	0.635
VC	0.226	0.1105	0.566	1.009	0.067	0.128
weight	0.107	0.052	0.269	0.479	0.032	0.061

M Acts for Mean Value; VC Means Coefficient of Variation in the Table
From Table 1, the greater the coefficient of variation, the greater the weight of the influence factors. The weight of water quality factor in Zhoukou section was the highest, reaching 0.479, indicating that water quality was the key factor affecting the ecological base flow value in this section. The weights of discharge factor and precipitation factor were 0.269 and 0.107 respectively, indicating that they were also important factors affecting, but the impact of discharge was greater than that of precipitation. The weight of economic, evaporation and water consumption factors was 0.061, 0.052 and 0.032, which revealed that the three factors had little effect on the ecological base flow value of the river reach and were secondary factors.

M Acts for Mean Value; VC Means Coefficient of Variation in the Table
By Table 2, water quality was also key factor affecting the ecological base flow value of Shenqiu section, with weight of 0.365; The weights of discharge and precipitation factor was 0.301 and 0.141 respectively, which were important factors, but the influence of the former was greater than that of the latter. The weights of economic, evaporation and water consumption factors were accordingly 0.081, 0.058 and 0.054, indicating that the three factors had little influence on ecological base flow value.

Table 2. Coefficient of variation and weight of influence factors in Shenqiu segment

year	C1	C2	C3	C4	C5	C6
2008	0.843	1.044	0.097	0	0.939	0.480
2009	0.995	0.998	0.129	0	0.945	0.486
2010	1.304	0.947	0.057	3.001	1.008	0.497
2011	1.173	0.817	0.052	0	0.927	0.500
2012	1.784	0.891	0.068	1.667	0.873	0.554
2013	0.852	0.910	0.152	1.667	0.931	0.575
2014	0.775	0.889	0.273	1.667	0.822	0.589
2015	1.044	0.978	0.170	0.882	0.942	0.584
2016	0.663	0.932	0.259	1.740	0.913	0.593
2017	0.842	1.060	0.084	1.463	0.857	0.590
2018	1.036	1.133	0.101	1.261	0.849	0.596
2019	1.388	1.220	0.237	1	0.953	0.723
2020	1.122	1.128	0.253	1	1.121	0.729
2021	1.069	1.052	0.057	1	1.170	0.771
M	1.064	1	0.142	1.168	0.946	0.59
VC	0.262	0.108	0.559	0.678	0.100	0.151
weight	0.141	0.058	0.301	0.365	0.054	0.081

3.2 Value of Ecological Base Flow in Different Years in the Middle Reaches

The ecological base flow value of Zhoukou and Shenqiu section in 2021 were 3.4759 and 1.2630 billion yuan respectively [15]. According to formula (10)–(12), the ecological base flow value of these two sections in 2008–2020 could be obtained, as shown in Table 3.

Table 3. Ecological base flow value in different years in the two reaches billion yuan

Year	2008	2009	2010	2011	2012	2013	2014
Zhoukou	1.2486	1.3910	0.9975	0.8939	1.1503	1.5496	2.4157
Shenqiu	0.7755	0.9099	2.2044	0.6765	1.6818	1.7806	2.1561
year	2015	2016	2017	2018	2019	2020	average
Zhoukou	3.9411	3.9710	3.5244	4.1953	4.2571	3.9558	2.6379
Shenqiu	1.5053	2.1330	1.4643	1.4703	1.9065	1.9126	1.5600

According to Table 3, the average value of ecological base flow in Zhoukou segment from 2008 to 2021 was 2.6379 billion yuan, which varied from 0.8939 to 4.2571 billion yuan, with the minimum and maximum value appearing in 2011 and 2019 respectively. The value of Shenqiu section varied from 0.6765 to 2.2044 billion yuan and average value was 1.560 billion yuan, with the minimum value in 2011 and the maximum value in 2010.

4 Discussion

To some extent, the value of water resources depends on its quantity, that is, the scarcity and abundance of water resources. The more resources, the lower their value [16]. In general, the value of ecological base flow in Zhoukou section is greater than that in Shenqiu section. The runoff and precipitation of the latter are more abundant than that of the former by compared with data of the two stations, but the value of Shenqiu Section is lower, indicating that the ecological base flow also accords with the characteristics of resource scarcity.

Water quality is the key factor affecting ecological base flow value in same reach. Han Yapping [17] believed that the ecological base flow discharged from the reservoir should first meet the water quality requirements of the cross section. The research from Zhang Yunlin [18] showed that human predatory exploitation and utilization of water body was the main reason for the decline of ecological quality in the middle and lower reaches of the Yangtze River. Tian Liufang [19] pointed out that the irrational exploitation and utilization of river runoff by human beings was the key to endangering thc ecological base flow and its value.

The water quality of Zhoukou Section from 2008 to 2014 and Shenqiu section from 2008 to 2009 and 2011, which did not reach the set target, and annual rate of compliance was 0, which made the value of ecological base flow in the year was relatively lower, indicating that among the water quality, precipitation and flow factors of the same section, the influence of water quality was the most important on ecological base flow value.

From the studies of Hu Xiaoyang [20], the main factors affecting water quality of the River was that agricultural non-point source pollution in villages and towns on both sides of the river and non-standard discharge of waste water from some enterprises. The use value and purpose of water resources largely depend on water quality. Shen Danhui [21] suggested that enterprises and drainage households should be taxed on pollution by the loss degree of drainage quality to river value, so as to regulate human behavior, reduce the soil erosion and non-point source pollution in the basin, improve the water quality and mitigate the impact of unreasonable behaviors on ecological base flow and its value.

5 Conclusions

Based on the data of rainfall, runoff, evaporation, water quality compliance rate, per capita water consumption and per capita GDP of Zhoukou and Shenqiu sections in the middle reaches of Shaying River from 2008 to 2021, the impact factor matrix of ecological base flow was constructed, and the value of ecological base flow in different years was determined by coefficient of variation method combined with the value of ecological base flow in 2021, and their change characteristics were analyzed. The results were as follows:

(1) The ecological base flow value of Zhoukou section fluctuated between 0.8939–4.2571 billion yuan, and that of Shenqiu section between 0.6765 and 2.2044 billion yuan; the variation coefficient and weight of water quality factors were both the largest in the two reaches, which was the key factors affecting value of ecological base flow. The weight of discharge and precipitation were in the middle, which were important factors affecting the value of ecological base flow, but the effect of discharge was greater than that of precipitation. The weight of economy, evaporation and water consumption were smaller, indicating their effect on ecological base flow value of river were smaller.
(2) The ecological base flow value of the two reaches were affected by amount of resources in some extent. The annual discharge and precipitation of Shenqiu Reach were greater than that of Zhoukou Reach, but its value of ecological base flow was smaller than that of the latter on the whole, indicating that ecological base flow had the characteristic of scarcity of water resources.

To sum up, it can be seen that coefficient of variation method is suitable for temporal and spatial change analysis of ecological base flow value of river, which can avoid the interference of human activities, find out relatively objective main factors affecting ecological base flow and its value, promote in-depth development of its guarantee work, and ensure ecological security of basins.

References

1. Mu, W., Yu, F., Li, C., et al.: Differences of river ecological base flow in concept and evaluation method and its influence. China Rural Water Hydropower (1), 90–94 (2015)
2. Hu, K., He, J., Zhao, J., et al.: Ecological base flow in Niya River Basin under climate change. Arid Land Geogr. **45**(5), 1472–1480 (2022)
3. Li, Q., Si, C.: An empirical study on the development level of inclusive finance in Guangxi Province based on coefficient of variation method. J. Reg. Financ. Res. **6**, 79–83 (2017)
4. Du, J.: Research of fuzzy comprehensive water quality assessment model based on different methods. Yellow River **37**(12), 69–73 (2015)
5. Jiang, F., Zhang, B.: Analysis and evaluation of water resources carrying capacity in Jiuquan based on AHP and variation coefficient method. Water Saving Irrig. **9**, 94–99 (2023)
6. Zhao, M.: Research on measuring the development level of inclusive finance in China based on coefficient of variation method. J. Jilin Financ. Res.(1), 14–18, 37 (2019)
7. Zhang, W., Gu, X.: Comprehensive evaluation of wind farms using variation coefficient method and technique for order preference by similarity to ideal solution. Power Syst. Technol. **38**(10), 2741–2746 (2014)
8. Luo, Z.: Interaction mechanism and distributed model of social economy - water - ecology in the Shaying River Basin. Doctoral dissertation of Zhengzhou University (2019)
9. Xu, M.: Temporal and spatial variation of ecological base flow value in Guanzhong Section of Weihe River mainstream. Master thesis of Xi'an University of Technology (2017)
10. Li, H., Yue, S., Zhao, Y.: Research progress in value of ecological basic flow. J. Econ. Water Resour. **33**(4), 6–9, 14 (2015)
11. Cheng, B., Li, H.: Research on agricultural ecological compensation based on river ecological base flow. J. Nat. Resour. **32**(12), 2055–2064 (2017)
12. Li, H., Huang, W., Yue, S.: The value calculation of ecological basic flow based in Fuzzy mathematical model. Ecol. Econ. **32**(5), 186–190 (2016)

13. Kou, S.: Study on the influence of large-scale water conservancy projects on flood season. Tech. Superv. Water Conserv. **5**, 87–92 (2020)
14. Wang, N., Li, Q., Zhao, L., et al.: Analysis in ship detection with high resolution SAR images based on variation coefficient method. Remote Sens. Technol. Appl. **32**(2), 305–314 (2017)
15. Wang, G., Li, Z., Yang, D., et al.: Estimation and analysis of ecological base flow value in main stream of Shaying River in Henan Section. Res. Soil Water Conserv. (2024). https://link.cnki.net/urlid/61.1272.P.20240611.1051.003
16. Jia, Y., Shen, J., Wang, H., et al.: Research on Fuzzy evaluation of water resources value in Lanzhou City. Yellow River **40**(9), 68–73 (2018)
17. Han, Y.: Study on ecological discharge of reservoir based on cross section water quality objectives. Water Resour. Power **39**(5), 50–53 (2021)
18. Zhang, Y., Zhang, Y., Qin, B., et al.: Evolution process and driving mechanism of lake ecological space in the middle and lower reaches of the Yangtze River. Environ. Sustain. Dev. **5**, 33–36 (2019)
19. Tian, L.: Analysis of basic flow variation trend in Jinghe River basin. Ground Water **37**(5), 153–155 (2015)
20. Hu, X., Xu, Y., Wang, H., et al.: Analysis of water quality change and influencing factors in important gate section of Shaying River. Water Power **44**(11), 1–5 (2018)
21. Shen, D.: Research on annual environmental economic accounting of water resources based on water quality value in Shanghai City. Master thesis of East China Normal University (2017)

Impact of Crop Cover on Soil Erosion

Tustha Srivalai[1], Siwawut Pattanasri[2], Paitoon Jaikla[3], and Sunisa Smittakorn[1](✉)

[1] Faculty of Engineering, Thammasat School of Engineering, Thammasat University, Klong Luang 12121, Pathum Thani, Thailand
ssunisa@engr.tu.ac.th
[2] Aurecon Consulting, Bangkok, Thailand
[3] Inter-Expert Construction, Nakhon Pathom, Thailand

Abstract. Soil erosion has made quite an impact on the agricultural industry because the nutrient can be lost along with the erosion. Moreover, those sediment can be carried away resulting in the blockage of the river, lake or even the dam. As such that it can lead to the reduction of the drainage capacity and eventually to the flood. Generally, there are 3 approaches to prevent the soil erosion; namely soil and water conservation with plant, soil and water conservation with mechanic and lastly, soil and water conservation with cultivation. It should be noted that such methods can only be reduced the damages causing by the soil erosion. The Soil and water conservation with plant can be implemented by cover cropping, crop rotation, contouring, strip cropping, and diversion terrace.

In this study, a cover cropping was proposed as a solution to the soil erosion. The plant itself will act as a shield from the impact of the raindrop which would then prevent the detachment of the soil particle. Spanish Shawl and Amarillo Peanut were selected as the cover crop. Poorly graded sand (SP) was the soil sample in this study. Studies were performed on a mini-plot of the width of 0.2 m with 2 lengths of 1.6 and 2.4 m. Slopes of the plot were 5, 20, and 35°. There were 2 rainfall intensities of 63 and 117 mm/h applied onto the plot. All experiments were done on 3 soil conditions; without any cover cropping, with Spanish Shawl and with Amarillo Peanut as a cover crop. The soil erosion would be considered only as the interrill erosion.

Results showed that there was 70 to 90% reduction of the soil erosion using cover cropping. Those two plants in this study can be easily obtained with relatively low-cost price. Their characteristics such as relatively deep roots, fast-growing, and low-maintenance are appealing as cover cropping. It is recommended to plant these crops in low-angle slope areas or any empty plots to prevent the soil erosion. Lastly, further analysis was performed to get a relationship of the interrill erosion. It was found that in case of no cover cropping and Amarillo Peanut; slope played the major role in soil erosion while the runoff rate had the least effect on the soil erosion. However, the runoff rate had most effect to the soil erosion and the rainfall intensity offered the opposite effect for Spanish Shawl.

Keywords: Interrill erosion · Soil erosion · Mini-plot experiments

C. Yuan (Ed.): ICRER 2024, ESE, pp. 68–79, 2025.
https://doi.org/10.1007/978-3-031-93295-3_6

1 Introduction

One issue that exists on a global scale is soil erosion [1]. The consequences of soil erosion include land degradation that reduces fertility, and its productivity sedimentation that blocks open channels which eventually leading to increased flooding. Many countries have problems with loss soil nutrient from soil erosion [2, 3]. Erosion threatens agricultural output and water quality by reducing soil fertility and causing major environmental issues. Beyond biological and environmental damage, soil erosion has wide-ranging effects. The economic implications include decreased fishing and farming production, fewer tourism opportunities, higher water purification costs, and deteriorated public health [4]. According to estimations from the land development department, the ministry of agriculture and cooperatives, Thailand, lost nutrients such as nitrogen, phosphorus and potassium, from soil erosion cost Thailand roughly 240 million dollars in 2020 [5]. Soil erosion also affected the water treatment plant facilities in northeastern provinces of Thailand. It was discovered that operating costs at the Khon Kaen Municipal Treatment facility were significantly higher during a wet season than they were during a dry one. [6] According to a 2003 assessment on environmental management in Thailand's 12,000 km^2 Phong watershed, the rate of soil erosion is 0.12 mm per year. Due to the soil erosion, sediment built up in the Ubolratana dam until 1990 at a rate of 37.50 million tons, or roughly 1.5 million tons each year since 1965 [7].

Two main processes of soil erosion is described; soil detachment and sediment transport that contribute to overland soil erosion during rainfall [8]. The process called interrill erosion begins when a raindrop strikes the ground, which is followed by soil dissociation and sediment transport to the rill [9, 10]. Once detached soil occurred from the interrill erosion, the sediment being carried through the tiny channels on the ground. Raindrops are considered to be a primary factor in soil separation [11]. A study found that, when slope and vegetation cover were held constant, sediment detachment caused by raindrop impact was related to the intensity of the rainfall. [12]. A mini-plot experiment has been an important tool to investigate the interrill erosion because it can eliminate the effect of rill erosion. Many studies have been focused on establishing interrill erosion equation via small scale experiments. To further understand the interril erosion process, a 4-m-long plot was tested. The analysis revealed that the soil erosion rate increased in the higher area before decreasing in the bottom section. It is assumed that a raindrop initiates the initial control of erosion and that sediment is later delivered in the lower portion [13]. It was suggested that additional research be done to better understand how the interrill process is impacted by soil types, slope, and rainfall intensity. In WEPP model based on mini-plot experiments, an interrill erosion equation was first proposed to estimate interrill erosion rates [14];

$$D_i = K_i I^2 S_f \tag{1}$$

where D_i is interrill erosion rate (kg m^{-2} s^{-1}), K_i is interrill erodibility, I is rainfall intensity (m s^{-1}) and S_f is slope factor that can be calculating using the following equation [15];

$$S_f = 1.05 - 0.85\exp(-4\sin\theta) \tag{2}$$

where θ is the slope gradient (°).

A modified interrill erosion equation was proposed by adding the factor of overland flow into the original equation which can be shown as a separate runoff parameter as the follow equation [16–18];

$$D_i = K_i RIS_f \tag{3}$$

where R is runoff rate (m s^{-1}). Another interrill erosion equation based on mini-plot experiment was proposed. Experiments were tested on the same soil type but with five levels of slope lengths, five levels of slope gradient, and five levels of rainfall intensities. The improved interrill erosion equation was then transformed as follows [19];

$$D_i = K_i RI^{0.22} S_f L^{-0.25} \tag{4}$$

where L is slope length (m). One of the major causes of soil erosion is land use change especially the cropland expansion. According to a study, Southeast Asia, South America, and Sub-Saharan Africa have the largest potential threats from soil erosion rates. [20]. Land management such as the cultivated crop, tillage method, crop cover application and crop canopy cover can be applied to reduce the soil erosion [21]. Synthetic and natural geotextiles have been studied to prevent the soil loss [22]. One study examined on the interaction between the geotextiles and the vegetation growth. The findings indicated that the investigated geotextiles may have an impact on vegetation growth and may, over time, provide inadequate soil protection [23].

Vegetation cover has been used to reduce water and soil erosion in many cases such as in olive orchards [24]. It was discovered that, regardless of the type of crop, olive groves with no cover crops caused higher rates of soil erosion than those with cover crops. As a result, it should be highlighted that sections of bare soil and areas of olive trees without a cover crop displayed identical results in terms of soil erosion. In reality, the interception of rainfall by olive trees was quite low.

The plant's ability to intercept rain has two major effects, the most significant of which is a decrease in the erosive force of falling raindrops. Moreover, it lessens the amount of water that reaches the soil's surface. Hence, one of the fundamental strategies for reducing soil erosion across all land use types can be accomplished by altering land use and increasing ground coverage [25]. Plant cover has been applied to reduce the soil erosion. Fern-mulching was shown to be an effective treatment after the prescribed burning by reductions in runoff coefficients and soil losses by 70–80% [26]. Studies on various types of plant showed significant efficiency to reduce soil loss [24, 27]. A two-year study of the impact of different types of fodder on runoff and soil erosion showed that the runoff and sediment yields were related to the rainfall intensity and the soil cover extension with a threshold when fodder covers more than 40% of plot area [28]. The main objective of this study was to evaluate the impact of vegetation cover on soil erosion including the factors of rainfall intensity, slope length, slope gradient. The vegetation cover would serve as a barrier to prevent interrill erosion caused by the impact of rainfall. Three different scenarios for the mini-plot studies will be presented: one with no vegetation cover, one with Amarillo Peanuts, and one with Spanish Shawl.

2 Materials and Methods

2.1 Study Site, Soil and Plant Types

Experiment setups had been performed at the Civil engineering laboratory in Thammasat University, Thailand. A mini-plot with 3 different slopes of 5, 20 and 65° filled with lateritic soils were tested under 2 rainfall intensities of 63 and 117 mm/h. The mini-plot was established at 2 lengths of 1.6 and 2.4 m. A simple rain simulator was constructed to get a designed raindrop size and distribution. Two types of plants which were Amarillo Peanut and with Spanish Shawl were selected as a cover cropping based on these criteria; it must germinate quickly, require minimum maintenance other than watering and occasionally feeding. Finally, the soil erosion experiments were performed under three conditions; without any plant cover, with Amarillo Peanut and with Spanish Shawl as a crop cover.

2.2 Experimental Setups

This work was separated into two parts; the preparation of the rainfall simulator, the setting up a mini-plot for plant cultivation and plant the crop prior to the experiment. After all the preparations were done, several erosion experiments were performed under these conditions;

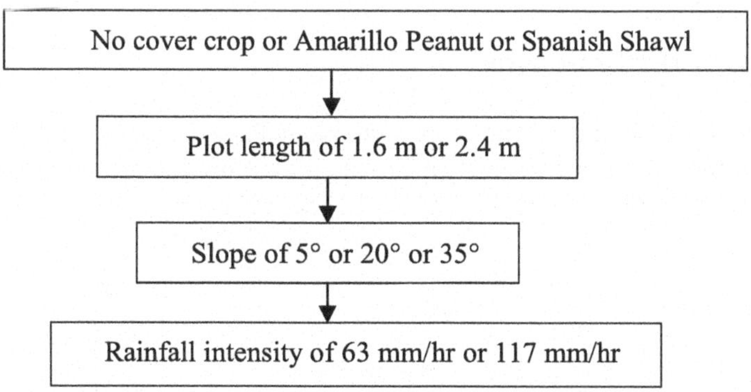

A simple rainfall simulator was constructed about 12.5 m above the mini-plot to create precipitation with 2 intensities. The raindrop size and its distribution were measured with a simple low-cost method [29] by setting up a 1*1 m^2 plate covered with 2 mm of flours beneath the rainfall simulator. After the sprinkler was turned on, the raindrops made a spherical shape impression on the flour plate. The diameter of raindrops was measured using a Vernier caliper. As for the rainfall distribution, several rain gauges were placed all over the tested area to collect the precipitation sprayed from the rainfall simulator.

Lateritic soil was collected from Chonburi province to be tested for soil erosion experiment. Three sets of mini-plot were investigated; one without any crop cover, one with Amarillo Peanut and the last one with Spanish Shawl. Spanish Shawl or the scientific name of *Heterocentron elegans* (Schltdl.) Kuntze is native plant to the Central America countries such as Mexico and Honduras. Amarillo Peanut or *arachis pintoi* or the scientific name of *Arachis pintoi Krapov. & W.C. Greg* is native crop to Brazil. Both plants can grow well in warm weather, are relatively drought resistance, and can rapidly grow with minimal maintenance. Spanish Shawl and Amarillo Peanut grow well in warm and humid climate such as Thailand and can easily find in Thailand at a reasonable low cost.

All mini-plot were constructed with the plastic frame measuring 1.6 m or 2.4 m (length) × 0.2 m (width) × 0.165 m (depth) and were prepared to the soil density of about 1,100 kg/m^3. Both plants were grown beforehand to ensure that they can grow uniformly across the entire plot. Once the plant was ready, the mini-plot was transported to the test stand where the slope can be adjusted to three different degree. Two rainfall intensities were generated to test the soil erosion. It should be noted that even though the rainfall's intensity varied, it could be divided into two ranges: one with an average lower range of 66 mm/h and one with an average upper range of 117 mm/h. The depth of the surface runoff was measured at the upstream, midstream, and downstream ends of the plot. Runoff and interrill wash sediments were intended to be gathered at the end of the plot. For the erosion analysis, the collected sediments were oven-dried and weighed.

3 Results and Discussions

The interrill erosion rate from each case were plotted against the rainfall intensity, the slope factor, the runoff rate and the slope length according to the improving interrill erosion equation as shown in Fig. 1, Fig. 2 and Fig. 3;

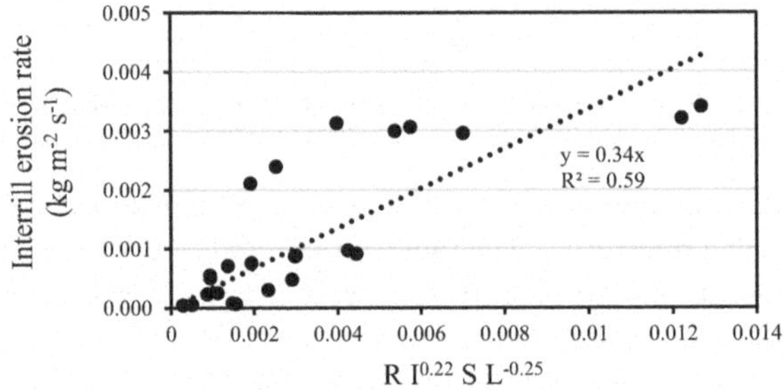

Fig. 1. The interrill erosion equation for no crop cover

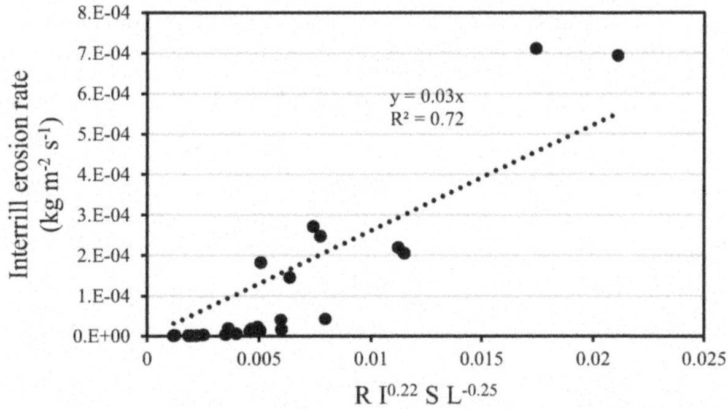

Fig. 2. The interrill erosion equation for Spanish Shawl

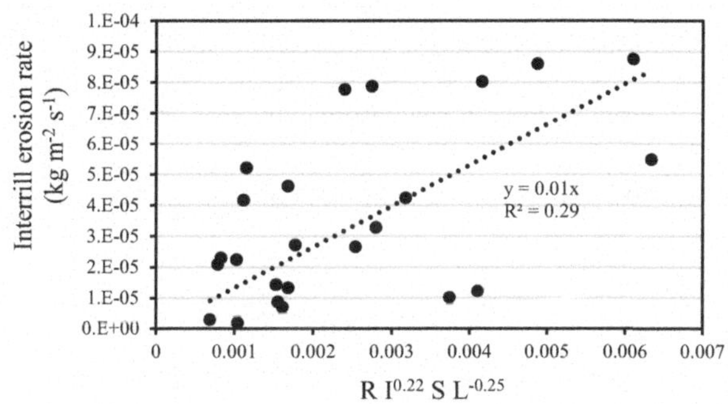

Fig. 3. The interrill erosion equation for Amarillo Peanut

Interrill erodibility from mini-plot experiment for laterite with 3 different slopes, 2 rainfall intensities and 2 plot length were evaluated for all 3 cases. The coefficient of determination (R^2) for no vegetation cover case, Spanish Shawl and Amarillo Peanut are 0.59, 0.72 and 0.29, respectively. Only the R^2 of Spanish Shawl indicated that the interrill erosion rate fit quite well with the improving interrill erosion equation. This suggests that the link between the interrill erosion rate and its influencing factors was not effectively described by the improving interrill erosion.

To find the factors affecting the interrill erosion by applying backward stepwise regression, the subsequent stage of this analysis would then be carried out independently for each of the three examples. For each case, parameters in the experiments namely rainfall intensity, slope factor, runoff rate and slope length will be tested to verify their statistically significant. It should be noted that all data from the experiments were used in the statistical analysis. Results showing P-value for all three cases as shown;

All of the parameters in Table 1 had P values greater than 0.05 which indicated that all parameters in Table 1 have no statistically significant to the soil erosion. As a result, until the statistically significant difference between parameters can be demonstrated, non-significant variables will be eliminated from the analysis one at a time. The final analysis yields the results as shown;

Table 1. P-value from regression analysis

Parameter	No Cover	Spanish Shawl	Amarillo Peanut
I	0.470	0.458	0.455
I^2	0.487	0.507	0.507
Log I	0.498	0.419	0.434
S_f	0.000	0.006	0.074
R	0.879	0.123	0.278
R^2	0.837	0.108	0.221
Log R	0.849	0.625	0.389
L	0.000	0.226	0.118

Table 2. Final P-value from regression analysis

Parameter	No Cover	Spanish Shawl	Amarillo Peanut
I	0.000		
S_f	0.000	0.000	0.009
R		0.000	
R^2	0.015	0.000	0.024
L	0.000	0.002	0.000

After parameters with the highest P-value and ones with P-value larger than 0.05 had been removed, the remaining significant parameters are shown in the Table 2. From Table 2, it shows that parameters affecting the interrill erosion are the rainfall intensity, the slope factor, the square of runoff rate, and the slope length for the case of no vegetation. For Spanish Shawl and Amarillo Peanut only the slope factor, the runoff rate and the slope length are the major factor influencing the interrill erosion rate. It should be emphasized that when vegetation covers the soil surface, regardless of the type of vegetation, rainfall intensity has no influence on interrill erosion. Another proof that vegetation cover prevented soil erosion can be found in here.

For the statistical analysis, the relationship between the interrill erosion and the variables can be written in Eq. (5), (6) and (7) for no vegetation cover, Spanish Shawl, and Amarillo Peanut, respectively.

$$\ln D_i = 5.48S_f + 2.01L + 31,109I - 28.98R^2 - 16.09 \tag{5}$$

$$\ln D_i = 4.42S_f - 2.04L + 89.01R - 204.81R^2 - 16.07 \tag{6}$$

$$\ln D_i = 1.95S_f - 1.71L + 248.34R^2 - 11.42 \tag{7}$$

Predicted interrill erosion using Eq. (5) for no vegetation are plotted against the interrill erosion from the experiment as shown in Fig. 4;

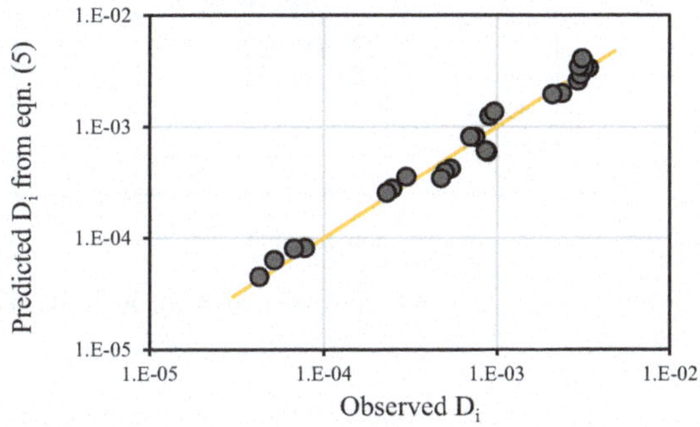

Fig. 4. Predicted vs. observed interrill erosion for no vegetation cover

Comparison between the predicted interrill erosion for no vegetation cover and the interrill erosion from the experiment, ratio between predicted interrill erosion to the observed one is 0.67–1.4.

Predicted interrill erosion using Eq. (6) for Spanish Shawl are plotted against the interrill erosion from the experiment as shown in Fig. 5;

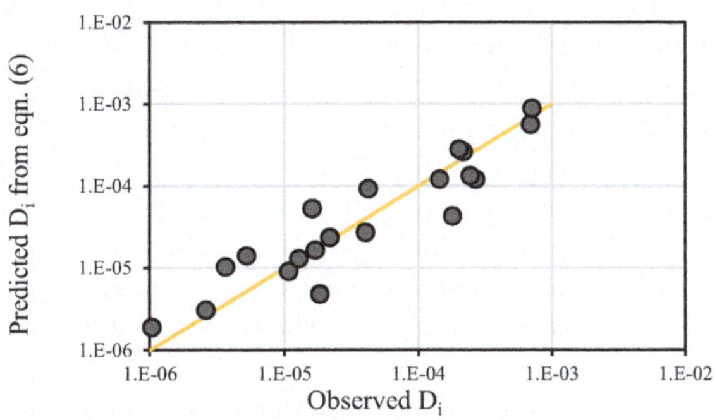

Fig. 5. Predicted vs. observed interrill erosion for Spanish Shawl

The ratio of projected interrill erosion for Spanish Shawl to observed interrill erosion is 0.24–3.28 when compared to the interrill erosion from the experiment.

Predicted interrill erosion using Eq. (7) for Amarillo Peanut are plotted against the interrill erosion from the experiment as shown in Fig. 6;

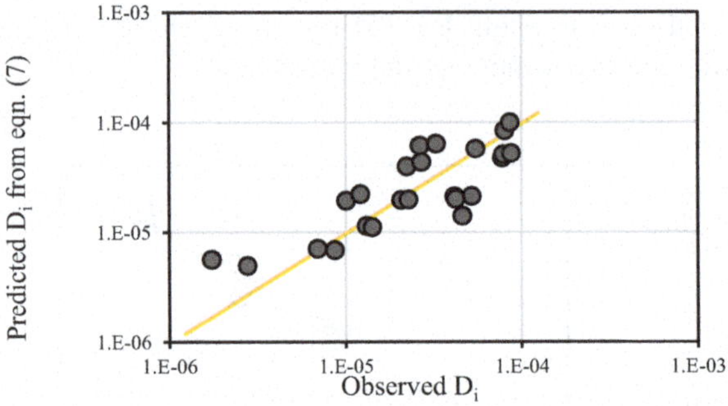

Fig. 6. Predicted vs. observed interrill erosion for Amarillo Peanut

When compared to the interrill erosion from the experiment, the ratio of the expected interrill erosion for the Amarillo Peanut to the observed interrill erosion is 0.30–3.25. When the soil is covered with vegetation, regardless of the type of vegetation, it is demonstrated by comparing Eq. (5) to Eqs. (6) and (7) that the intensity of the rainfall has no bearing on the interrill erosion.

Both the slope factor and the slope length play significant roles in interrill erosion. For no vegetation cover case, both the slope factor and the slope length are positively affected the interrill erosion. The slope factor, as opposed to the slope length, has a favorable impact on interrill erosion for the Spanish Shawl and Amarillo Peanut, though. Indicating, the longer travelling distance, it is less likely to gain more erosion when there are vegetation cover. Effect of slope factor agrees with previous work which indicated that slope gradient has a significant effect on the sediment load [30]. The slope length had a favorable impact in the case of no crop cover, but it had a reverse impact on the Spanish Shawl and Amaiollo Peanut. One study found that the rates of soil erosion initially increased and then dropped as the slope length increased [13]. In contrast to this study, where soil samples were obtained at the end of the plot, their approach involved collecting sediment in each section along a 4-m-long test. The decrease in soil erosion along the length of the plot suggested that transport predominated in the lower section while erosion predominated in the top section. The structure of the crop may slow down transport when there is vegetation present.

In all three scenarios, the runoff rate and the interrill erosion have a quadratic connection. While there is a negative correlation between the Spanish Shawl and no vegetative cover, there is a positive correlation for the Amarillo Peanut.

From previous works, there are several studies showed the relationship between the rainfall intensity and the soil erosion [31, 32]. In this study, the rainfall had a direct, favorable impact on soil erosion. However, for the Spanish Shawl and Amarillo Peanut, it is obviously shown that the rainfall intensity have no effect on the interrill erosion due to the leaf coverage over the soil surface. For slope lengths of 1.6 m and 2.4 m, the percentage of the area covered by Spanish Shawl is around 92.7 and 94.2%, respectively. For slope lengths of 1.6 m and 2.4 m, the percentage of the area covered by Amarillo Peanut is around 74.68 and 77.6%, respectively. Both plants were set up at the same time and given the same care, such as exposure to sunlight. The characteristics of each plant account for the variability in the lead coverage area.

Given the quantity of soil erosion that happened in the case with no plant cover as the baseline, the ability to prevent soil erosion can be compared. The soil erosion estimating from the case of Spanish Shawl and Amarillo Peanut may be compared as shown in Fig. 7. It is demonstrated that the rate of soil erosion in Spanish Shawl and Amarillo Peanut, respectively, is only approximately 7.78 and 12.12. An obvious improvement is demonstrated in comparison when there is no soil surface protection against the rainfall.

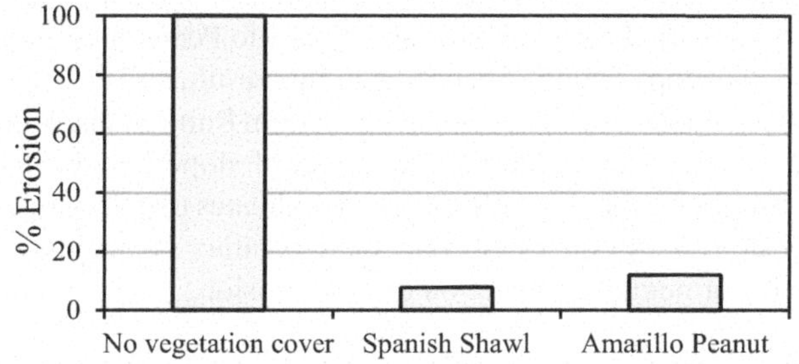

Fig. 7. Comparison of erosion percentage for all cases

Considering the efficacy of erosion reduction for each plant, it is shown that Spanish Shawl can reduce soil erosion for 92.21% while Amarillo Peanut can prevent soil erosion for 87.87% as shown in Fig. 8. The difference in leaf coverage area between those two plants may be responsible for a modest variation in soil protection. This finding is consistent with other research on the relationship between plant functional diversity and vegetation cover, which shows that the amount of vegetation cover has the biggest impact on interrill erosion. Sediment yield decreased by 83% on a plot with 60% plant cover [33].

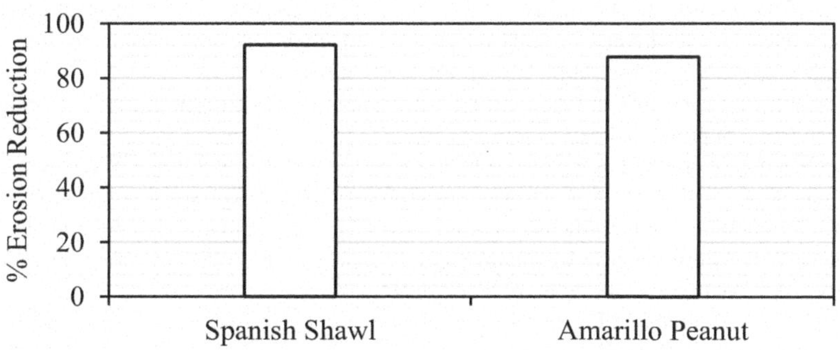

Fig. 8. Percent erosion reduction for Spanish Shawl and Amarillo Peanut

4 Conclusion

Factors affected the interrill erosion such as rainfall intensity, plot length, slope factor, and vegetation cover had been studied. Series of experiment were performed in a mini-plot set. The test subjects included two varieties of low-maintenance perennial plants and a case with no plant cover. Two distinct rainfall intensities on two different plot lengths with three slope gradients were tested. Results from this experiment did not fit well with the previous improved interrill erosion equation. After a set of statistical analysis, it showed that the slope gradient has a direct impact on the soil erosion in all cases. The slope length and soil erosion in the case of bare soil have a positive relationship. On the other hand, both the Spanish Shawl and Amarillo Peanut cases demonstrated the negative effects of the slope length. According to this result, soil erosion would become more severe over greater distances devoid of vegetation. Rainfall has little impact on soil erosion with vegetation cover, much like the effect of slope length. In all three cases, there is a similar pattern for the slope effect, which indicates that the slope has a favorable impact on soil erosion. As for the runoff rate, there is rather inconclusive result whether it has either positive or negative impact on the soil erosion.

Acknowledgement. Financial support for this project had been provided by Faculty of Engineering, Thammasat School of Engineering, Thammasat University, Thailand.

References

1. Borrelli, P., et al.: An assessment of the global impact of 21st century land use change on soil erosion. Nat. Commun. **8** (2017)
2. Peri, P.L., Lasagno, R.G., Chartier, M., Roig, F., Rosas, Y.M., Pastur, G.M.: Soil erosion rate and nutrient loss in Rangeland of Southern Patagonia. In: Imperiled: The Encyclopedia of Conservation, pp. 102–110 (2022)
3. Chapungu, L., Nhamo, G., Dube, K., Chikodzi, D.: Soil erosion in the savanna biome national park of South Africa. Phys. Chem. Earth Part A/B/C **130** (2023)
4. Chicas, S.D., Omine, K., Ford, J.B.: Identifying erosion hotspots and assessing communities' perspectives on the drivers, underlying causes and impact of soil erosion in Toledo's Rio Grande watershed: Belize. Appl. Geogr. **68**, 57–67 (2016)
5. Status of soil erosion in Thailand, Land development department, Ministry of agriculture and cooperatives, September 2020
6. Khon Kaen University (KKU): Environment management of the Phong Watershed (2003)
7. Sthiannopkao, S., Takizawa, S., Homewong, J., Wirojanagud, W.: Soil erosion and its impacts on water treatment in the northeastern provinces of Thailand. Environ. Int. **33**, 706–711 (2007)
8. Ellison, W.D.: Some effects of raindrops and surface flow on soil erosion and infiltration. Trans. - Am. Geophys. Union **26**, 415–429 (1945)
9. Young, R.A., Wiersma, J.L.: The role of rainfall impact in soil detachment and transport. Water Resour. Res. **9**(6), 1629–1636 (1973)
10. Owoputi, L.O., Stolte, W.J.: Soil detachment in the physically based soil erosion process: a review. Trans. ASAE 38(4) (1995)
11. Nouhou-Bako, A., Darboux, F., James, F., Lucas, C.: Rainfall erosivity in interrill areas: insights about the choice of an erosive factor. Catenam **180**, 21–31 (2019)
12. Gabet, E.J., Dunne, T.: Sediment detachment by rain power. Water Resour. Res. **39**(1) (2003)

13. Zhang, X.C., Nearing, M.A., Garbrecht, J.D.: Gaining insights into interrill erosion processes using rare earth element tracers. Geoderma **299**, 63–72 (2017)
14. Elliot, W.J., Liebenow, A.M., Laflen, J.M., Kohl, K.D.: A compendium of soil erodibility data from WEPP cropland soil field erodibility experiment 1987 and 1988, NSERL report no.3, The Ohio State University, and USDA Agricultural Research Service (1989)
15. Elliot, W.J., Laflen, J.M., Kohl, K.D.: Effect of soil properties on soil erodibility, paper no. 89-2150. Transactions of the ASAE, St. Joseph, MI (1989)
16. Truman, C.C., Bradford, J.M.: Laboratory determination of interrill soil erodibility. Soil Sci. Soc. Am. J. **59**, 519–526 (1995)
17. Kinnell, P.I.A.: Runoff as a factor influencing experimentally determined interrill erodibilities. Aust. J. Soil Res. **31**, 333–342 (1993)
18. Flanagan, D.C., Nearing, M.A.: USDA-Water erosion prediction project: Technical documentation, NSERL Rep no.10, National soil erosion research lab, West Lafayette, IN (1995)
19. Zhang, F.-B., Wang, Z.-L., Yang, M.-Y.: Validating and improving interrill erosion equations. PLOS ONE **9**(2) (2014)
20. Borrelli, P., et al.: An assessment of the global impact of 21st century land use change on soil erosion. Nat. Commun. (2013)
21. Matthews, F., Verstraeten, G., Borrelli, P., Panagos, P.: A field parcel-oriented approach to evaluate the crop cover-management factor and time-distributed erosion risk in Europe. Int. Soil Water Conserv. Res. **11**, 43–59 (2023)
22. Bhattachaaryya, R., Smets, I., Fullen, M.A., Poesen, J., Booth, C.A.: Effectiveness of geotextiles in reducing runoff and soil loss: a synthesis. CATENA **81**, 184–195 (2010)
23. Alvarez-Mozos, J., et al.: Evaluation of erosion control geotextiles on steep slopes, part 2: influence on the establishment and growth of vegetation. CATENA **121**, 195–203 (2014)
24. Beniaich, A., et al.: Spontaneous vegetation as an alternative to cover crops in olive orchards reduces water erosion and improves soil physical properties under tropical conditions. Agric. Water Manag. **279** (2023)
25. Nunes, A.N., Almeida, A.C., Coelho, C.: Impact of land use and over type on runoff and soil erosion in a marginal area of Portugal. Appl. Geogr. **31**, 687–699 (2011)
26. Carra, B.G., Bombino, G., Lucas-Borja, M.E., Plaza-Alvarez, P.A., D'Agostino, D., Zema, D.A.: Prescribed fire and soil mulching with fern in Mediterranean forests: effects on surface runoff and erosion. Ecol. Eng. **176** (2022)
27. Cerda, A., et al.: The role of plant species on runoff and soil erosion in a Mediterranean shrubland. Sci. Total Environ. **799** (2021)
28. Hai, A.P.H., et al.: Impact of fodder cover on runoff and soil erosion at plot scale in a cultivated catchment of North Vietnam. Geoderma 177–178 (2012)
29. Mazon, J., Vinas, M.: A low-cost experiment for determining raindrop size distribution. Weather **68**, 49–52 (2013)
30. Balacco, G.: The interrill erosion for a sandy loam soil. Int. J. Sediment Res. **28**, 329–337 (2013)
31. Nearing, M.A., Yin, S., Borrelli, P., Polyakov, V.O.: Rainfall erosivity: an historical review. CATENA **157**, 357–362 (2017)
32. Gabet, E.J., Dunne, T.: Sediment detachment by rain power Water Resour. Res. **39**(1) (2003)
33. Martin, C., Pohl, M., Alewell, C., Korner, C., Rixen, C.: Interrill erosion at disturbed alpine sites: effects of plant functional diversity and vegetation cover. Basic Appl. Ecol. **11**, 619–626 (2010)

Sustainable Resource Management and Circular Economy

Addressing Food Loss and Waste: A Study of Milk Loss in Pasteurized Milk Factory

Thirapha Praradorntham[1,2] and Orathai Chavalparit[3,4](✉)

[1] International Program in Hazardous Substance and Environmental Management, Graduate School, Chulalongkorn University, Bangkok 10330, Thailand
[2] Center of Excellence on Hazardous Substance Management (HSM), Chulalongkorn University, Bangkok 10330, Thailand
[3] Department of Environmental and Sustainable Engineering, Faculty of Engineering, Chulalongkorn University, Bangkok 10330, Thailand
orathai.c@chula.ac.th
[4] Research Unit of Environmental Management and Sustainable Industry, Faculty of Engineering, Chulalongkorn University, Bangkok 10330, Thailand

Abstract. Milk loss in dairy factories in Thailand is a complicated problem that results from several steps in the dairy manufacturing process. Further losses may result from processing inefficiencies, including spills, broken equipment, and antiquated technologies. The cleaning process, filling, and packaging method frequently cause spoilage, exacerbating the problem. Significant technical expenditures, sophisticated processing, and packaging solutions are all part of the effort to minimize spoilage and shrink these losses. Furthermore, this study investigates creative waste valorization techniques to turn wasted milk into valuable byproducts that support sustainable manufacturing and the circular economy. It is crucial to accurately quantify milk loss and identify its specific occurrence points within the production process. Implementing the waste hierarchy and valorization strategies mitigates adverse environmental consequences and provides financial benefits, including cost savings and additional revenue. When taken as a whole, these ambitions aim to increase sustainability, lower milk loss, and boost the profitability of Thailand's dairy industry.

Keywords: Milk Loss · Pasteurized Milk Factory · Waste Valorization

1 Introduction

Food loss and waste (FLW) is a critical global issue that undermines efforts toward long-term sustainability, impacting social, environmental, and economic dimensions. With over 820 million people facing starvation and more than two billion experiencing food insecurity, the paradox of discarding one billion metric tons of food annually highlights the severity of this problem [1, 2]. FLW constitutes the largest portion of global municipal solid waste, with the East Asia and Pacific region [3], including Thailand, contributing significantly to this issue. Thailand alone generates over 7.1 million metric tons of

C. Yuan (Ed.): ICRER 2024, ESE, pp. 83–94, 2025.
https://doi.org/10.1007/978-3-031-93295-3_7

food waste each year, driven by rising population growth and consumption patterns [4]. The wastage of food also involves the wastage of crucial resources such as land, water, energy, and labor, further intensifying its impact on sustainability. In response, the United Nations introduced Sustainable Development Goal (SDG) 12, specifically target 12.3, which aims to halve global food waste by 2030 and reduce losses along the entire food supply chain. Food loss and waste can occur at various stages, including production, storage, processing, distribution, and consumption. Understanding where and why these losses occur is essential for addressing the problem effectively. While food loss tends to occur earlier in the supply chain, food waste is more prevalent at the retail and consumer levels. Given the complexity of food supply chains and the varied definitions of FLW, this study aligns with the FAO's terminology to ensure clarity in addressing SDG 12.3.

This paper focuses on milk, particularly pasteurized milk, which falls under the FAO's category of animal products with a high potential for FLW reduction. Milk plays a vital role in human nutrition across all life stages, and its global production is projected to increase, especially in developing countries like Thailand [5, 6]. Despite its importance, Thailand's dairy sector, particularly small- and medium-sized enterprises, lacks comprehensive data on milk loss and waste. This lack of information hampers efforts to balance supply and demand in the milk supply chain, leading to inefficiencies and sustainability issues. The dairy supply chain, although seemingly straightforward compared to other food supply chains, involves several stages (Fig. 1):

- *Feed production* is the first step in the dairy supply chain that makes crops grow for feeding milch animals.
- *Milk production* is the step that houses the milch animals, including giving food to collect their milk.
- *Milk transportation* is the activity that transports milk from farm to dairy plant or processing.
- *Packaging* is the activity that designs to keep dairy products fresh, clean, and wholesome and done by the dairy processor.
- *Distribution* refers to the action of the processor delivering milk products to retailers, wholesalers, schools, and others.
- *Retail* is selling milk and dairy products to the public or consumers.
- *Consumers* utilize milk and dairy products as essential nutrient diets.

Each stage adds value to the product but also presents opportunities for loss. Processing is particularly crucial, as milk is highly perishable and requires timely and efficient handling to prevent spoilage. Therefore, promoting sustainable practices in dairy processing and management is essential for minimizing FLW and achieving SDG 12.3.

The objectives of this study are twofold: first, to quantify the milk loss within the pasteurized milk production line in Thailand, and second, to analyze the causes of loss along the dairy supply chain. By identifying critical points of inefficiency, this study aims to support the development of possible options to reduce milk loss and enhance sustainability in the dairy industry.

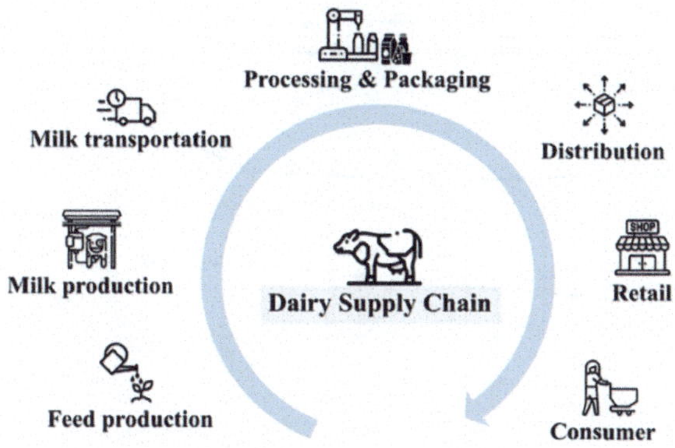

Fig. 1. Dairy supply chain. Adapted from [7].

2 Dairy Sector in Thailand

In Thailand, growing food industries make production more than 30 million metric tons per year, and it rose to 40 million metric tons during 2017 [8]. The food industry, which includes the dairy industry, has a notable role in driving Thailand's economic growth and had 9,516 registered factories in 2019 [9]. Cow milk is mainly Thailand's milk production, and it had roughly 1.3 million tons of milk production with 724,901 dairy cow heads in 2020. In Fig. 2, the preliminary data of raw milk production and dairy cow heads (estimated on 1st January of the year) continue growing as better practices and standards in dairy farm management, including the government policy that emphasizes improving the quality of raw milk to meet the standards and safety for consumers [10, 11].

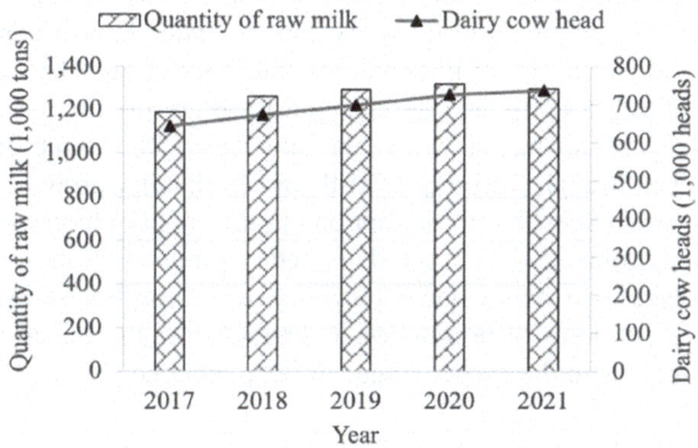

Fig. 2. Thailand's milk production and dairy cow heads (2017–2021) [11].

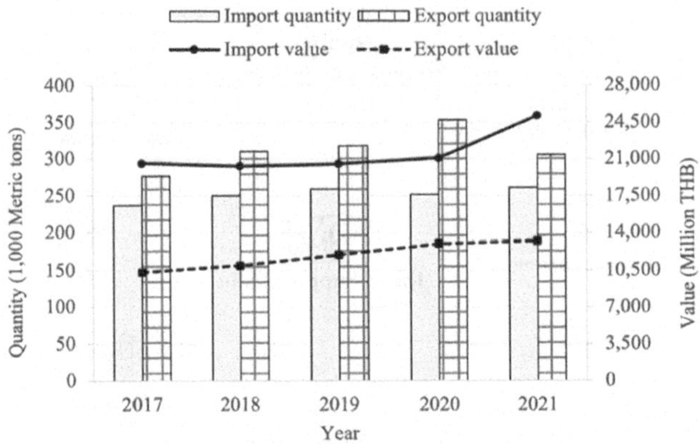

Fig. 3. Thailand's import/export quantity and value of milk and dairy products (2017–2021) [12, 13].

At present, Thailand has 154 registered dairy factories [9, 14]. UHT milk - the process of milk sterilization through ultra-high temperature - drinking yogurt, yogurt, and buttermilk account for the importance of Thailand's milk and dairy product exports [13, 14]. In 2019 although the volume of exported milk and dairy products was 317,529 metric tons in the approximate value of 12 billion THB, the number of imported milk and dairy products was 259,257 metric tons that came to 21 billion THB (Fig. 3) [13]. Large amounts of milk and dairy product imports are milk powder, cream, and whey. Skim milk powder can be made into various products such as ready-to-drink milk, condensed milk, bread, ice cream, yogurt, sweetened condensed milk, candy, and chocolate. Preventing milk loss can lead to a reduction in the need for importing milk and dairy products. Indeed, saving more avoidable milk loss can result in increased milk and dairy product production without changing the input volume. It means to use resource efficiency and promote sustainable development in the dairy supply chain. As demonstrated in Fig. 2 and 3, the milk production, the milk cowherds, the import, and the export continued growing because of larger herds and the demand for dairy products. OECD/FAO also reports that the dairy sector has shown remarkable resilience to the pandemic of the COVID-19 virus even though fresh liquid milk and fresh dairy products are perishable foods which made them sensitive to disruptions in the supply chain. Nevertheless, the pandemic did not significantly affect the dairy sector. In general, there were no severe surpluses and shortages worldwide due to the dairy sector's ability to respond quickly to the pandemic by adjustments of production and packaging; moreover, it alleviated the initially extreme impacts in the earlier stages of the pandemic [15].

3 Methods

After reviewing the literature on the dairy sector in the previous section, the lack of information highlights gaps in data regarding Thailand's milk loss and waste status. However, the growth of milk output and the dairy industry leads to an essential need for sustainable performances, including the effective and efficient prevention in changing milk into milk and dairy products, to minimize the pressure on the environment, society, and economy. Hence, a quantitative and qualitative approach is applied in this study to clarify milk loss in the dairy industry based on pasteurized milk products. The main steps of the research method framework contained: case selection, data collection, and data analysis. Research diagram of this study shows in Fig. 4.

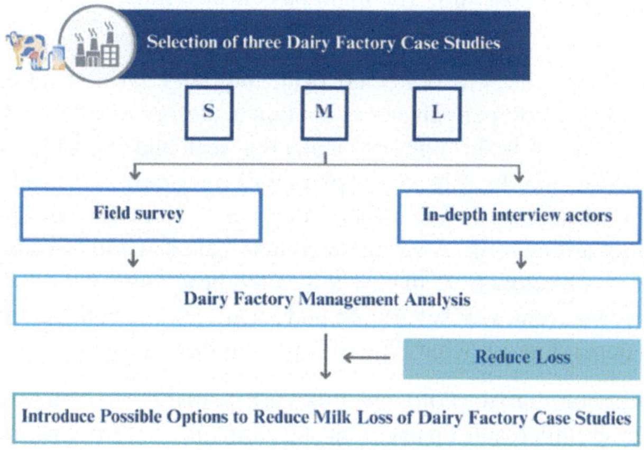

Fig. 4. Research diagram.

The criteria for selecting factories for milk losses quantification were factory capital intensive, factory size, location, time, and cost; thus, this study chose a factory placed in Bangkok and the vicinity province of Thailand to determine the losses and answer the study objectives. For milk loss investigation, the site survey was conducted by an open-end questionnaire as a research tool in an in-depth interview (face-to-face). The questionnaire primarily concentrated on identifying sources of milk loss and examining current management practices to mitigate it. The interviewee who answered the question was the person who knew or connected with the environment, production lines, and milk loss management (e.g., safety officer, quality assurance, and environmentalist). Existing milk loss data was gathered during the site survey. The study also collected data on milk loss management for perceiving the destination of the loss. Data on milk loss in both quantity and management were examined by flow analysis and descriptive method to gain a deeper understanding of the results.

4 Results and Discussions

This section delineates survey responses and then compiles milk loss flow in each factory. However, there are two routes for ready-to-drink milk in Thailand: the commercial market and the Supplementary Food Project (Milk) of Schools or other school milk programs. The establishment of the project is to support the Thai dairy industry and attempt to solve the issue of malnutrition and nutritional deficiencies in children: young at an early stage. This program also contributes an outlet to locally produced milk by giving

200 ml of milk to the youths in primary and kindergarten schools for 260 days a year. For this analysis, before conducting the site survey, the study required submitting request letters for permission to conduct research at dairy factories. Three factories accepted these requests including two in Saraburi province (two factories) and Suphanburi province (one factory), central Thailand, as detailed in Table 1. They all had the same primary pasteurized milk-making processes following the Good Manufacturing Practice (GMP) by Thai Ministry of Public Health. The main pasteurized milk processing steps of three factories were:

Milk Received: Raw milk was collected from the dairy farm and transferred to the milk processing factory/company in a cool container. Before receiving the raw milk, the company tested the raw milk through physical, chemical, and microbiological testing. If all raw milk meets the quality criteria, the plant will weigh the approved milk and pump it into refrigerated storage tanks for further processing. If not, the dairy processing plant will reject that unqualified milk. A refrigerated tank kept raw milk in cold condition at a temperature of 8 °C or below 6 °C in case it awaited processing but not over 48 h. Later action from the storage tank was separation and fortification, but these steps may or may not be available depending on what type of milk and dairy product.

Separating: This step spun the cold raw milk via a centrifuge to isolate the weightier fat milk from the lighter milk resulting in cream and skim milk or remixing the cream and the remaining milk to the desired fat content milk product. The separator additionally cleaned the raw milk by removing dirt, some bacteria, and other sedi-ments that might have been present. Then, food additives or essential ingredients such as vitamins, minerals, fatty acids, probiotics, and phytoconstituents were added in the fortifying step to decline nutritional deficiencies; moreover, the fortification was a complementary tool to convey the essential and functional substances to consumers.

Thermization: This step can be called primary heating, using heat to warm the milk for 15 min at 57 to 68 °C to kill harmful bacteria that pose serious health risks and leave good bacteria in the product. Other benefits include maintaining the milk's structure and flavor were unaffected by the low temperature.

Homogenization: Homogenizing milk has assured that milk fat is spread evenly throughout the milk. In addition, it prevented the milk fat from floating to the surface.

Pasteurization: Thermal processing is used in dairy factories to make products safe for consumption while extending their shelf life by destroying pathogenic microorganisms and diminishing enzymes in milk. The pasteurizer heated the milk in the pipeline continuously for at least 15 s at a temperature above 72 °C, later cooling down to 5 °C or below. Furthermore, the heated milk traveled through a long pipe with a length and diameter that allowed the liquid to pass from side to side within the exact period requirement. If the milk's temperature falls below the standard, the sensor at the end of the pipe redirects the milk back to the input for reprocessing.

Filling and Packaging: The milk was fully prepared and waited in a storage tank for the next step, filling and packaging. The milk was pumped and sealed into various containers, including pouches, plastic bottles, and others, with an aseptic system to prevent recontamination. A date label was printed on each container as it moved through the production line, indicating the production and expiration date or how long the milk stays fresh. After that, packaging managed the containers into boxes, cartons, or palettes that helped with easy storage and transportation.

Storage and Distribution: The packaged milk was chilled at 4 °C or below 8 °C and was ready for shipment to shops and clients. It was transported in refrigerated trailers to distribution warehouses and then to retail locations, where it was kept in refrigerated display cases until it was taken.

Table 1. The details of dairy factories.

Factory	Size	Province	Interviewee	Production Capacity (tonnes/day)	%loss [tonnes/day]
A	S	Suphanburi	Mechanic and Electrician	6	0.78 [0.0475]
B	M	Saraburi	Quality assurance	35	21.61 [9.65]
C	L	Saraburi	Safety officer and Environmentalist	667	0.22 [1.5]

S: Small factory (capital intensive < 50 million THB).
M: Medium factory (capital intensive 50 – 200 million THB).
L: Large factory (capital intensive > 200 million THB).

4.1 Milk Loss and its Management in Each Factory

Factory A

A small-sized facility in Suphanburi province, efficiently manages twice-daily milk intakes from large and small farmers in the morning and late afternoon. The large farmer contributed 10 tons of milk every two days, which were transported directly in cool containers to the factory's tank. In contrast, the small farmer delivered 900 kg of milk daily, which requires cooling before transfer to storage tank. The factory produced six tons of pasteurized milk daily, packaged in 200 ml plastic bags for the school program. However, despite their efficiency, the factory experienced some milk losses: 20 kg during receiving, another 20 kg during pasteurization, 5 kg during filling, and 2.5 kg during packaging.

Milk losses were identified at various stages of the production process. During the receiving and pasteurization stages, the primary source of milk loss was traced in the pipeline. At the filling stage, milk loss occurred due to outdated milk bagging technology. Additionally, during the packaging stage, losses were attributed to the inefficient sealing of bags. The cumulative milk loss was calculated to be 47.5 kg per day. To manage this waste, the factory drains the lost milk into three constructed ponds around the facility.

Factory B

Factory B, a medium-sized local dairy cooperative in Saraburi province, Thailand, processed approximately 95 tons of raw milk daily, sourced from various milk receiving centers. This milk was predominantly utilized for producing pasteurized and UHT milk. However, during the production of pasteurized milk, some losses occurred. Specifically, around 1.05 tons of milk was lost in the pipes during the receiving stage, an additional 0.5 tons during pasteurization, and 8.1 tons during packaging. Notably, some of the milk lost during packaging could potentially be recovered through reprocessing.

The factory experienced substantial milk losses, primarily during the packaging stage, due to inefficiencies in bag sealing and frequent tears. These losses total 18,900 kg per day; however, a significant portion can be reprocessed into pasteurized milk products. Additional losses happened in the pipeline during milk reception and pasteurization. Overall, the total daily milk loss amounted to 9,650 kg. These losses were managed using an aerated lagoon system.

Factory C

Factory C, a large-sized pasteurized milk processing plant located in Saraburi province, Thailand, processed pasteurized milk and received milk shipments twice daily from dairy cooperatives nationwide. Significant milk losses were identified during the filling and packaging stages, amounting to 1,500 kg per day. An additional 100 kg of losses stemmed from expired or nearly expired products, including yogurt, flavored pasteurized milk, and other dairy items. To effectively manage these challenges, the factory explored bioenergy generation as a sustainable solution for utilizing these losses efficiently. Nonetheless, the efficiency of biogas production was found to be surprisingly low.

Examining the pattern of milk loss across three different factories exposes the stage at which loss occurs (Fig. 5) and their overall impact, indicating areas that might use improvement. According to the study, there are five main reasons why milk is lost: (1) cleaning and flushing manufacturing lines; (2) obsolete machinery and technology; (3) packaging issues; (4) equipment failures; and (5) expired products, similar to [16] and [17]. While Factory C continuously services the commercial market, Factory A and B provide pasteurized milk to the school program, operating 260 days a year. Factory A experiences the highest milk loss in the pipeline during receiving and pasteurization, likely due to the batch-based operation of the production process; rather than continuous 24-h running, milk is wasted during cleaning cycles. Each time the process stops for cleaning, any milk remaining in the pipes is discarded, resulting in a loss. Factory B has the highest milk loss, especially during the packaging phase, meaning serious equipment or material problems that require immediate attention. Factory C loses milk mostly during the filling and packaging, highlighting the necessity of routine maintenance or preventive maintenance schedule and possible equipment improvement. Packaging issues show up as a recurring concern in all factories, though to varying degrees. The factories must capitalize on their milk losses, representing a significant missed opportunity. Implementing effective waste reduction and value optimization strategies, such as improving biogas production efficiency in Factory C to convert milk loss and waste into a sustainable energy source, reduces environmental impacts while recovering economic value from lost milk.

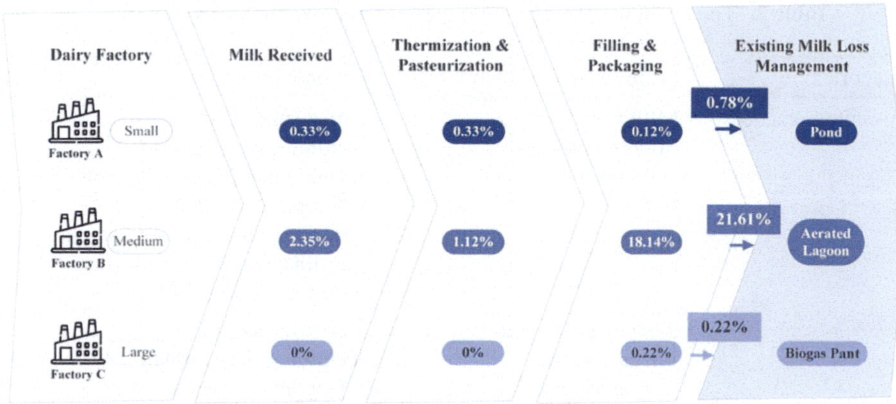

Fig. 5. The occurrence of milk loss in production stage.

4.2 The Possible Option to Loss/Waste Minimization and Valorization

In line with the waste management hierarchy, prevention is the top priority, followed by reuse, recycling, recovery, and disposal [1]. Consequently, the first action in managing loss/waste effectively for the three factories is to prevent it from arising by cooperating with clean technology practice and lean manufacturing. Milk loss and waste that cannot be managed with preventive methods should be subjected to waste valorization.

Waste valorization is the process of diverting waste materials into more valuable energy or goods, improving economic advantages, and lessening the environmental impact. Valorizing milk loss and waste in the context of dairy factories can diminish the negative consequences of disposal and generate extra income. The waste management hierarchy coordinates with waste valorization by converting loss and waste into beneficial resources. The manufacturers advance up the hierarchy, encouraging resource efficiency and sustainability. Regarding to the analysis in this study and literature reviews the possible options to minimize and valorize milk loss and waste displays in Table 2.

Table 2. The possible options to minimize and valorize milk loss and waste.

Categories	Stage (source of FLW)	Possible options
Prevention practices	**Milk received** (loss in pipeline)	**Improvement of technology/equipment** • Upgrading pipeline systems • Adopting a continuous milk production setup with automated in-line cleaning to reduce milk loss during cleaning cycles
	Thermization and Pasteurization (loss in pipeline)	**Improvement of technology/equipment** • Upgrading pipeline systems • Investing modern thermization and pasteurization machine • Upgrading pipeline systems
	Filling and packaging	**Improvement of technology/equipment** • Investing modern filling and packaging machinery with improved sealing capabilities • Expand reprocessing capacity for salvaging milk loss **Improvement of preventative maintenance** • Changing equipment components on a regular basis to avoid failure • Regular equipment servicing • Schedule routine maintenance to prevent machine breakdowns
Others	**Milk loss and waste, dairy wastewater**	**Valorization of milk loss and waste** • Biofuels • Biomethane [18] • Biohydrogen (H_2) [19, 20] • Biodiesel [21, 22] • Biodiesel and Bioethanol [23] • Bioplastic • poly-beta-hydroxy-butyrate, PHB [20] • Enzymes • Lipases [24]

5 Conclusion

The analysis of three dairy factories in Thailand highlights milk losses at various production stages due to pipeline inefficiencies, outdated equipment, packaging issues, and machine breakdowns. Collectively, these factories process large volumes of milk daily but face common challenges. Factory A, a small operation supplying milk to a school program, loses about 47.5 kg daily (approximately 0.78% of its production), mainly during receiving and pasteurization due to pipeline problems and outdated filling and sealing equipment. Factory B, a medium-sized cooperative, experiences substantial losses totaling 9,650 kg per day (over 20% of its production), predominantly during the packaging stage because of inefficient sealing and frequent bag tearing, alongside pipeline losses; some of this lost milk is reprocessed. Due to filling and packaging problems brought on by machine failures and expired and nearly expired goods, Factory C, the sizable facility catering to the commercial sector, loses over 1,600 kg per day.

The effective utilization of milk loss and waste is significantly hindered by the low efficiency of biogas production processes that rely on these materials, as the organic composition of milk loss and waste often leads to challenges such as imbalanced nutrient ratios, suboptimal microbial activity, and limited methane yield, making it less viable as a sustainable energy source without further optimization. All the factories need to capitalize on their milk losses, indicating a significant opportunity to implement waste reduction and value optimization strategies through technology upgrades, process improvements, and sustainable waste management practices aligned with the hierarchy.

References

1. FAO: The State of Food and Agriculture 2019. Moving forward on food loss and waste reduction (2019)
2. FAO: Waste and the Right to Adequate Food: Making the Connection. Rome: 48 (2018)
3. Kaza, S., Yao, L., Bhada-Tata, P., Van Woerden, F.: What a Waste 2.0: A Global Snapshot of Solid Waste Management to 2050. World Bank Publications (2018)
4. Dung, T.N.B., Sen, B., Chen, C.-C., Kumar, G., Lin, C.-Y.: Food waste to bioenergy via anaerobic processes. Energy Proc. **61**, 307–312 (2014)
5. Mojarab Soufiyan, M., Aghbashlo, M.: Application of exergy analysis to the dairy industry: a case study of yogurt drink production plant. Food Bioprod. Process. **101**, 118–131 (2017)
6. OECD/FAO: OECD-FAO Agricultural Outlook 2018–2027. OECD Publishing, Paris/FAO, Rome (2018)
7. Rodriguez-Enriquez, C., Alor-Hernández, G., Sanchez-Ramirez, C., Córtes-Robles, G.: Supply chain knowledge management: a linked data-based approach using SKOS. Dyna **82**, 27–35 (2015)
8. OIE: Industrial Economic Status Report for 2018 and Outlook for 2019 (2018). http://www.oie.go.th/en/academic/Industrial%20Economic%20Status%20Report%20for%202018%20and%20Outlook%20for%202019. Accessed 17 Jan 2014
9. DIW: Factory Statistics (2020). https://www.diw.go.th/hawk/content.php?mode=Knowledge. Accessed 13 Jan 2020
10. MOAC and OAE: Agricultural Economic Information by Commodities 2020. Agricultural Statistics Number 402 (2021). https://www.oae.go.th/assets/portals/1/files/jounal/2564/commodity2563.pdf. Accessed 15 June 2024
11. MOAC and OIE: Agricultural Statistics of Thailand 2020. Agricultural Statistics Number 401 (2021). http://www.oae.go.th/assets/portals/1/files/jounal/2564/yearbook2563.pdf. Accessed 15 June 2024
12. MOAC and OAE: Situation of major agricultural products and trends in 2021 (2020). https://www.oae.go.th/assets/portals/1/files/jounal/2564/trend2564.pdf. Accessed 15 June 2024

13. OAE and CAI: Thailand Foreign Agricultural Trade Statistics 2020. Agricultural Statistics Number 403 (2021). https://www.oae.go.th/assets/portals/1/files/jounal/2564/trade2563.pdf. Accessed 15 June 2024
14. Bureau of Agricultural Economic Research: Situation of world and Thai dairy cattle in 2019 (2019). http://dairydevelopmentprogram.weebly.com/blog-3615363436193660361736263640358б/-2562. Accessed 15 June 2024
15. OECD/FAO: OECD-FAO Agricultural Outlook 2021–2030. OECD Publishing, Paris/FAO, Rome (2021)
16. Tóth, K., Borbély, C., Nagy, B., Szabó-Szentgróti, G., Szabó-Szentgróti, E.: Measurement of food losses in a Hungarian dairy processing plant. Foods **10**(2), 229 (2021)
17. Kazancoglu, Y., Ozkan-Ozen, Y.D., Ozbiltekin, M.: Minimizing losses in milk supply chain with sustainability: an example from an emerging economy. Resour. Conserv. Recycl. **139**, 270–279 (2018)
18. Kozłowski, K., et al.: Energetic and economic analysis of biogas plant with using the dairy industry waste. Energy **183**, 1023–1031 (2019)
19. Hellal, M.S., Abou-Taleb, E.M., Rashad, A.M., Hassan, G.K.: Boosting biohydrogen production from dairy wastewater via sludge immobilized beads incorporated with polyaniline nanoparticles. Biomass Bioenergy **162**, 106499 (2022)
20. Dinesh, G.H., et al.: Simultaneous biohydrogen (H2) and bioplastic (poly-β-hydroxybutyrate-PHB) productions under dark, photo, and subsequent dark and photo fermentation utilizing various wastes. Int. J. Hydrog. Energy **45**(10), 5840–5853 (2020)
21. Gupta, N., Manikandan, N.A., Pakshirajan, K.: Real-time lipid production and dairy wastewater treatment using Rhodococcus opacus in a bioreactor under fed-batch, continuous and continuous cell recycling modes for potential biodiesel application. Biofuels **9**(2), 239–245 (2018)
22. Kumar, A.K., et al.: Cultivation of Ascochloris sp. ADW007-enriched microalga in raw dairy wastewater for enhanced biomass and lipid productivity. Int. J. Environ. Sci. Technol. **16**(2), 943–954 (2019)
23. Chokshi, K., Pancha, I., Ghosh, A., Mishra, S.: Microalgal biomass generation by phycoremediation of dairy industry wastewater: an integrated approach towards sustainable biofuel production. Biores. Technol. **221**, 455–460 (2016)
24. Jaganmai, G., Rajeswari, J.: Optimization of lipase activity obtained from dairy industry wastes. Int. J. Eng. Res. Technol. (IJERT) **09**(11), 37–43 (2020)

Microwave-Assisted Synthesis and Characterization of Chitosan from Raw Tilapia (*Oreochromis Niloticus*) Scales

Tristan Roy Panaligan(✉), Jesuniño Aquino, Andrea Marie Alam, Jean Margaret Berroya, and Princess Lynelle Carysse Torres

Mapúa Malayan Colleges Laguna, 4025 Cabuyao, Laguna, Philippines
trlpanaligan@mcl.edu.ph

Abstract. This study investigates the synthesis of chitosan from tilapia fish scales using microwave-assisted methods—demineralization, deproteinization, and deacetylation—as a sustainable catalyst for biodiesel production. Microwave-assisted synthesis accelerated the extraction of chitosan from fish scales, reducing both processing times and energy consumption compared to conventional methods. Fourier Transform Infrared (FTIR) spectroscopy revealed characteristic absorption peaks at 3289.4 cm^{-1} and 3263.3 cm^{-1}, indicating O–H and N–H stretching, confirming the presence of amino and hydroxyl groups. Scanning Electron Microscopy (SEM) analysis at 1200x magnification showed a thick, non-porous structure with a rough surface and surface irregularities, consistent with previous studies on chitosan synthesized from fish scales.

Financial constraints during the time of doing the study limited the characterization of extracted chitin, impacting a full comparison of precursor and product morphologies. Despite these limitations, the findings emphasize the potential of microwave-assisted chitosan synthesis from tilapia fish scales as a promising and sustainable catalyst.

Keywords: Chitosan · Microwave-Assisted Technology · Raw Tilapia Fish Scales · FTIR · SEM

1 Introduction

Chitosan, derived from the deacetylation of chitin, is a biopolymer of growing significance due to its versatile applications across multiple industries. Chitin, the precursor to chitosan, is the second most abundant renewable natural polymer after cellulose and is sourced primarily from the exoskeletons of crustaceans, shrimp shells, fungi, and other natural materials [1]. This nitrogen-enriched polymer undergoes transformation into chitosan, which boasts unique properties such as biodegradability, biocompatibility, and high reactivity. These characteristics render it indispensable in fields ranging from medicine and agriculture to environmental sustainability and renewable energy [2, 3].

The applications of chitosan are extensive, spanning drug delivery, tissue engineering, food packaging, wastewater treatment, and the development of advanced biomaterials [3, 4]. Concerns regarding the effects on the environment and human health are

C. Yuan (Ed.): ICRER 2024, ESE, pp. 95–103, 2025.
https://doi.org/10.1007/978-3-031-93295-3_8

raised by the fact that traditional chitosan production methods, which include demineralization, deproteinization, and deacetylation, usually rely on prolonged reaction periods, significant energy consumption, and the use of hazardous substances. [3]. Of these, microwave-assisted methods have drawn interest because to their capacity to quicken chemical reactions, use less reagents, and improve energy efficiency—all of which are in line with the global movement toward sustainable practices [3, 4].

There are many benefits to producing chitosan using a microwave, such as increased yields, greater degrees of deacetylation (DDA), and smaller environmental impacts. A DDA of up to 85% is made possible by microwave technology, for example, outperforming traditional heating techniques while using fewer resources and producing less waste. [4]. This study covers important concerns about sustainability by using microwave radiation in the synthesis of chitosan, especially in the context of waste utilization. The study contributes to the circular economy by using tilapia fish scales, an abundant and underused biowaste, as the raw material for chitosan production.

This research directly contributes to two critical United Nations Sustainable Development Goals (SDGs). First, it supports SDG 12 (Responsible Consumption and Production) by transforming fish scale waste into a value-added product through sustainable synthesis methods, demonstrating resource efficiency and waste management. Second, it advances SDG 13 (Climate Action) by reducing energy consumption and chemical waste compared to conventional methods, offering a cleaner and more sustainable alternative for chitosan production.

This study specifically investigates the microwave-assisted synthesis of chitosan from tilapia fish scales through demineralization, deproteinization, and deacetylation. The synthesized chitosan is characterized using Fourier Transform Infrared (FTIR) spectroscopy and Scanning Electron Microscopy (SEM) to confirm its structural and functional properties. By demonstrating a scalable and sustainable approach to chitosan production, this research contributes with the aim to address global challenges in resource efficiency and environmental sustainability.

2 Literature Review

2.1 Chitin and Chitosan: Structures, Properties, and Functions

Chitin, a nitrogen-enriched natural polysaccharide, is the second most abundant biopolymer after cellulose. It is commonly extracted from crustacean shells, insect cuticles, and fish scales [5]. The structure of chitin comprises 2-acetamido-2-deoxy-D-glucopyranose units linked via β-(1,4) glycosidic bonds, which can form different crystalline allomorphs [6, 7].

Chitosan, derived from the deacetylation of chitin, exhibits significant functional versatility due to the presence of amino (-NH_2) and hydroxyl (-OH) groups. These functional groups enhance its reactivity, making chitosan an effective catalyst and a sought-after material for biomedical, agricultural, and environmental applications [4, 6, 8].

2.2 Fish Scales as a Resource for Chitosan

Fish scales, composed of collagen fibers, hydroxyapatite, and various calcium compounds, serve as a protective layer on fish skin. According to [9], these scales are typically discarded as waste during fish processing activities, contributing to environmental pollution. The fish processing industry generates approximately 18 to 30 million tons of waste annually, with fish scales accounting for around 4% of this volume. Despite their classification as waste, fish scales contain valuable components, including chitin, alkaloids, and phenols, making them a viable source for biotechnological applications [10]. With increasing interest in sustainable waste management, fish scales have been identified as a promising source of chitin, a precursor for chitosan synthesis [11].

Tilapia (*Oreochromis niloticus*) scales have garnered significant attention as a sustainable and underutilized source of chitosan, a biopolymer with extensive industrial and biomedical applications. [12] successfully extracted high-quality chitosan from Nile tilapia scales using a combination of acid and alkali treatments, underscoring its potential as a valuable biotechnological resource. Similarly, [13] demonstrated that fish scales can yield substantial amounts of chitosan when extraction conditions are carefully optimized.

Generally, fish scales contain 20–30% chitin, 30–40% proteins, 0–14% lipids, and 30–50% minerals [14]. The production of chitosan from fish scales not only addresses environmental concerns associated with fish waste disposal but also adds economic value through the conversion of waste into high-value products such as chitosan.

2.3 Conventional Methods for Chitosan Extraction

Traditional chitosan synthesis involves several steps: demineralization to remove calcium compounds, deproteinization to eliminate proteins, and deacetylation to convert chitin into chitosan. In the study by [14], they demonstrated the extraction of chitosan from fish scales using HCl and NaOH solutions, followed by heating for deacetylation. Similarly, [15] synthesized chitosan from Java fish scales using varying NaOH concentrations and temperatures, showing that deacetylation parameters significantly influence chitosan yield. While these conventional methods are effective, they often require prolonged reaction times and substantial chemical use, posing environmental and safety concerns [16].

2.4 Microwave-Assisted Chitosan Extraction Technology

Microwave technology has emerged as a promising alternative to conventional extraction methods, offering reduced reaction times, lower energy consumption, and decreased chemical use. The research of [5] demonstrated that microwave-assisted extraction of chitosan from shrimp shells achieved degrees of deacetylation (DD) exceeding 80% within 10 min, compared to several hours in conventional methods. Similarly, microwave heating improved demineralization and deproteinization efficiencies, removing over 95% of calcium carbonate ($CaCO_3$) in just a few minutes. Specifically, at 500 W and 650 W of microwave power, $CaCO_3$ removal increased from 91% to 96% and 94% to 98%, respectively, as demineralization time extended from 2 to 10 min. However, after 4 min, the removal efficiency plateaued, emphasizing the rapid effectiveness of microwave heating.

Using 500 W of microwave power for 8 min offers the ideal possible mix between practicality and efficiency, maximizing removal rates while reducing reaction time and energy consumption. These circumstances are especially beneficial for industrial applications, because shorter processing times can lead to higher throughput and significant cost savings.

In addition to demineralization efficiency, drying conditions also have a critical role in the preparation of biomaterials. In order to eliminate moisture and stabilize the material for further processing, fish scales were dried in an oven set at 80 °C for four hours, following the guidelines provided by [16] for shrimp leftovers. In order to ensure that all remaining moisture is removed and to get the material ready for successful integration with microwave technology in later stages, this drying stage is especially crucial.

Microwave-assisted technology has proven to be more efficient and environmentally friendly for several chemical reactions. The combined use of 8-min microwave-assisted demineralization at 500 W and a 4-h drying step at 80 °C represents a practical and scalable protocol for chitosan production, enhancing efficiency without compromising material quality [5].

3 Materials and Methods

3.1 Raw Material Preparation

The preparation of raw materials for chitosan synthesis began with the collection of approximately 1 kg of fish scales from local wet markets in different cities in Laguna, Philippines: San Pablo, Cabuyao, and Santa Rosa. Reagents and chemicals used to facilitate the chitosan synthesis, such as 1M HCl for demineralization, 1% vol. NaOH for deproteinization, and 30% vol. NaOH for deacetylation [8] were obtained from DKL Laboratory in Manila, Philippines. The synthesis of chitosan began with the preparation of tilapia fish scales. The scales were thoroughly washed, dried in a oven at 80 °C for 8 h, and then weighed using an analytical balance. This preparation set the stage for subsequent demineralization, deproteinization, and deacetylation processes with the use of microwave technology, which were adapted from [5, 14, 15] with some modifications. In Fig. 1, the process to obtain chitosan from Tilapia fish scales is described in summary.

3.2 Demineralization

The demineralization process involved treating 300 g of dried fish scales with 1500 mL of 1M HCl in a microwaveable polypropylene container for 8 min at 500 W. The demineralized scales were then filtered, washed with distilled water until neutral, and dried in an oven at 80 °C for 4 h.

3.3 Deproteinization

The process was conducted using a 1:1 (w/v) ratio of scales to NaOH, with microwave treatment at 500 W for 8 min. The scales were then filtered, washed until neutral, and dried at 80 °C for 4 h.

3.4 Deacetylation

Deacetylation converted chitin into chitosan by removing acetyl groups through treatment with a 50% vol. NaOH solution. The process was conducted with 30 g of chitin and 600 mL of NaOH, using microwave treatment at 500 W for 8 min. The deacetylated product was then washed with distilled water, dried at 80 °C for 4 h. A grinder/food processor was used to reduce the fish scales into a fine light brown powder. This was characterized using FTIR spectroscopy and SEM.

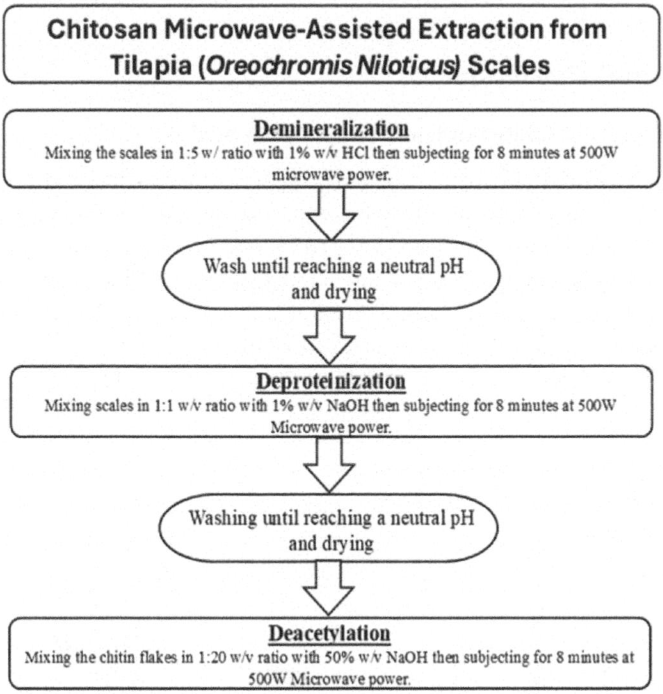

Fig. 1. Process Flow Summary of Microwave-Assisted Extraction of Chitosan from Tilapia Fish Scales

3.5 Fourier Transform Infrared (FTIR) Spectroscopy

FTIR was used to ensure the chemical structure and functional groups of chitosan made from raw tilapia scales met the requirements for its intended uses. The existence of functional groups, which are necessary for chitosan's reactivity and activity in a variety of applications, can be confirmed via FTIR analysis. The quality and structural integrity of the produced chitosan can be validated by comparing the spectrum results with existing data, confirming the viability of the extraction and synthesis procedure.

3.6 Scanning Electron Microscopy (SEM)

SEM imaging was used to evaluate the morphological and structural properties of chitosan made from tilapia fish scales that was employed with microwave technology. Key structural characteristics including porosity, thickness, and surface irregularities can be identified due to the SEM's precise insights into surface morphology. These characteristics have significance for determining the suitability and efficacy of chitosan, particularly when its physical properties have an impact on performance. Comparing the SEM results with the findings of previous research allows for a comprehensive evaluation of the synthesis process and its effectiveness in producing high-quality chitosan.

4 Results and Discussion

4.1 FTIR Results of Chitosan from Tilapia Fish Scales

Chitosan derived from raw tilapia scales was analyzed using FTIR spectroscopy, and the results are presented in Fig. 2. The absorption spectrum exhibited distinct peaks at 3289.4 cm^{-1}, 3263.3 cm^{-1}, 2935.3 cm^{-1}, 1636.3 cm^{-1}, 1420.1 cm^{-1}, 1397.8 cm^{-1}, and 1023.2 cm^{-1}. The bands at 3289.4 cm^{-1} and 3263.3 cm^{-1} indicated the presence of overlapping O–H and N–H stretching vibrations, characteristic of hydroxyl and amine groups, respectively. Additional peaks at 1420.1 cm^{-1} and 1397.8 cm^{-1} corresponded to O–H bending, further corroborating the presence of hydroxyl groups.

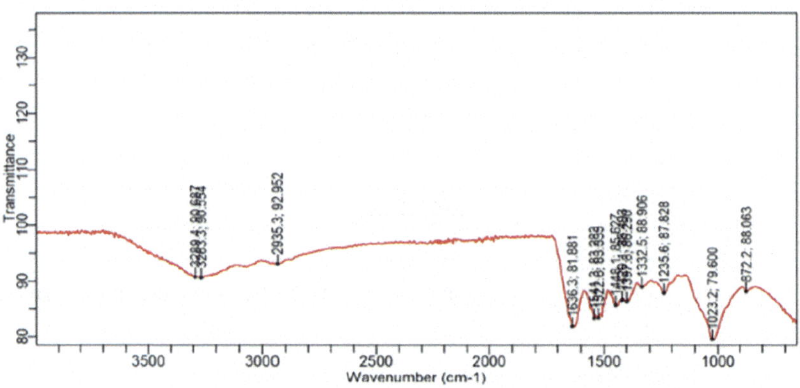

Fig. 2. FTIR Spectra of Chitosan from Tilapia Fish Scales

The peaks at 3289.4 cm^{-1} and 3263.3 cm^{-1} also suggest N–H stretching, indicative of aliphatic primary amines. This finding is further supported by bands in the ranges of 1650 cm^{-1} to 1580 cm^{-1} (N–H bending) and 1250 cm^{-1} to 1020 cm^{-1} (C–N stretching), both of which are characteristic of aliphatic amines. These major spectral patterns and indications are summarized in Table 1 and it align with those reported by [17], confirming the presence of amino and hydroxyl functional groups. Moreover, the observed peaks are consistent with those documented by [12], who synthesized chitosan from tilapia fish scales using a conventional method.

Table 1. FTIR Data and Indication Functional Groups of Chitosan

Wavenumber (cm^{-1})	Transmittance (%)	Functional Group
3289.4	90.687	O–H/N–H Stretch
3263.3	90.554	O–H/N–H Stretch
2935.3	92.952	C–H Stretch
1636.3	81.881	N–H Bend
1420.1	86.493	O–H Bend
1397.8	86.290	O–H Bend
1023.2	79.600	C–N Stretch

The similarity of absorption patterns validates the effectiveness of microwave-assisted synthesis in producing chitosan comparable to traditional methods. The presence of both amino and hydroxyl groups confirms the successful synthesis of chitosan from raw tilapia scales, demonstrating the potential of this innovative approach for efficient catalyst production.

4.2 SEM Micrograph of Chitosan from Tilapia Fish Scales

The morphology of chitosan synthesized using microwave technology was examined via SEM with the micrograph at 1200x magnification presented in Fig. 3.

Fig. 3. SEM Micrograph of Chitosan from Tilapia Fish Scales

The SEM analysis revealed a thick, nonporous structure, accompanied by a rough outer layer with several small fragments. This morphology aligns with findings from previous studies on chitosan synthesized from shrimp shells [5] and fish scales [13], both of which demonstrated dense and nonporous structures.

Additionally, the observed rough outer layer with small fragments is consistent with the morphology of chitosan derived from Java fish scale waste, as reported by [14], although their study employed lower magnifications of 50x and 200x. In contrast, chitosan synthesized from *Metapenaeus stebbingi* shells, as described by [18], exhibited flakey layers and porous regions, features absent in the current study's micrograph.

5 Conclusion

This study demonstrates the feasibility of synthesizing chitosan from tilapia (*Oreochromis niloticus*) fish scales using microwave-assisted methods. The approach significantly accelerates reaction times, achieving demineralization within 8 min at 500 W of microwave power, compared to the hours typically required in conventional methods. The resulting chitosan exhibited structural and functional characteristics, confirmed by FTIR and SEM analyses, which demonstrated the presence of critical functional groups such as amino (-NH_2) and hydroxyl (-OH) groups and a thick, nonporous morphology with a rough surface.

The findings emphasize the potential of repurposing fish waste, such as tilapia scales, into a value-added biopolymer while addressing environmental concerns associated with waste disposal. However, future research should focus on the comprehensive characterization of chitin prior to its conversion into chitosan to better understand the transformation process and optimize yields further.

Additionally, exploring the application of chitosan derived from tilapia fish scales in the conversion of waste vegetable oil to biodiesel offers a promising opportunity to integrate sustainable waste management practices with renewable energy production. Future studies could also explore broader applications of this chitosan to target additional UN SDGs, such as SDG 7 (Affordable and Clean Energy), thereby maximizing its environmental and socio-economic benefits.

Acknowledgment. The authors extend their heartfelt thanks to Mapúa Malayan Colleges Laguna and its constituent department, Mapúa Institute of Technology at Laguna, for providing academic resources and financial assistance for the publication of this study. Their significant contributions have been crucial to the successful completion of this research.

References

1. Zaeni, A., Safitri, E., Fuadah, B., Sudiana, I.N.: Microwave-assisted hydrolysis of chitosan from shrimp shell waste for glucosammine hydrochlorid production. J. Phys. Conf. Ser. **846**, 012011 (2017). https://doi.org/10.1088/1742-6596/846/1/012011. Author, F., Author, S.: Title of a proceedings paper. In: Editor, F., Editor, S. (eds.) CONFERENCE 2016. LNCS, vol. 9999, pp. 1–13. Springer, Heidelberg (2016)
2. Ji, J., et al.: Chemical modifications of chitosan and its applications. Polym. Plast. Technol. Eng. **53**(14), 1494–1505 (2014). https://doi.org/10.1080/03602559.2014.909486
3. Thambiliyagodage, C., Jayanetti, M., Mendis, A., Ekanayake, G., Liyanaarachchi, H., Vigneswaran, S.: Recent advances in chitosan-based applications—A review. Materials **16**(5), 2073 (2023). https://doi.org/10.3390/ma16052073

4. Aranaz, I., et al.: Chitosan: an overview of its properties and applications. Polymers (Basel) **13**(19), 3256 (2021). https://doi.org/10.3390/polym13193256
5. El Knidri, H., El Khalfaouy, R., Laajeb, A., Addaou, A., Lahsini, A.: Eco-friendly extraction and characterization of chitin and chitosan from the shrimp shell waste via microwave irradiation. Process. Saf. Environ. Prot. **104**, 395–405 (2016). https://doi.org/10.1016/j.psep.2016.09.020
6. Molina-Ramírez, C., Mazo, P., Zuluaga, R., Gañán, P., Álvarez-Caballero, J.: Characterization of chitosan extracted from fish scales of the Colombian endemic species Prochilodus Magdalenae as a novel source for antibacterial starch-based films. Polymers (Basel) **13**(13), 2079 (2021). https://doi.org/10.3390/polym13132079
7. Pellis, A., Guebitz, G.M., Nyanhongo, G.S.: Chitosan: sources, processing and modification techniques. Gels **8**(7), 393 (2022). https://doi.org/10.3390/gels8070393
8. Ooi, J., Lee, L.Y., Hiew, B.Y.Z., Thangalazhy-Gopakumar, S., Lim, S.S., Gan, S.: Assessment of fish scales waste as a low cost and eco-friendly adsorbent for removal of an azo dye: equilibrium, kinetic and thermodynamic studies. Bioresour. Technol. **245**, 656–664 (2017). https://doi.org/10.1016/j.biortech.2017.08.153
9. Qin, D., et al.: Development and application of fish scale wastes as versatile natural biomaterials. Chem. Eng. J. **428**, 131102 (2022). https://doi.org/10.1016/j.cej.2021.131102
10. Nur, R.M., Asy'ari, A.: The utilitation of fish scale waste as a chitosan. Agrikan: Jurnal Agribisnis Perikanan **13**(2), 269–273 (2020). https://doi.org/10.29239/j.agrikan.13.2.269-273
11. Boarin Alcalde, L., Graciano Fonseca, G.: Alkali process for chitin extraction and chitosan production from Nile tilapia (*Oreochromis niloticus*) scales. Lat. Am. J. Aquat. Res. **44**(4), 683–688 (2016). https://doi.org/10.3856/vol44-issue4-fulltext-3
12. Suneeta, K., Rath, P., Sri, H.K.A.: Chitosan from shrimp shell (Crangon crangon) and fish scales (Labeorohita): extraction and characterization. Afr. J. Biotechnol. **15**(24), 1258–1268 (2016). https://doi.org/10.5897/AJB2015.15138
13. Kumari, S., Rath, P.K.: Extraction and characterization of chitin and chitosan from (Labeo rohit) fish scales. Proc. Mater. Sci. **6**, 482–489 (2014). https://doi.org/10.1016/j.mspro.2014.07.062
14. Aichayawanich, S., Saengprapaitip, M.: Isolation and characterization of chitosan from fish scale waste. IOP Conf. Ser. Earth Environ. Sci. **301**(1), 012051 (2019). https://doi.org/10.1088/1755-1315/301/1/012051
15. Mohan, K., et al.: Green and eco-friendly approaches for the extraction of chitin and chitosan: a review. Carbohydr. Polym. **287**, 119349 (2022). https://doi.org/10.1016/j.carbpol.2022.119349
16. Santos, V.P., et al.: Recovery of chitin and chitosan from shrimp waste with microwave technique and versatile application. Arq. Inst. Biol. (Sao Paulo) **86** (2019). https://doi.org/10.1590/1808-1657000982018
17. Nandiyanto, A.B.D., Oktiani, R., Ragadhita, R.: How to read and interpret FTIR spectroscope of organic material. Indones. J. Sci. Technol. **4**(1), 97 (2019). https://doi.org/10.17509/ijost.v4i1.15806
18. Kucukgulmez, A., Celik, M., Yanar, Y., Sen, D., Polat, H., Kadak, A.E.: Physicochemical characterization of chitosan extracted from Metapenaeus stebbingi shells. Food Chem. **126**(3), 1144–1148 (2011). https://doi.org/10.1016/j.foodchem.2010.11.148

Circular Economy Transformation Through Intelligent Systems: A Systematic Review

Giang T. C. Tran[1,2(✉)], Thang Le Dinh[1], François Labelle[1], and Rahma Chouchane[1]

[1] Research Institute on SMEs, Université du Québec à Trois-Rivières, Québec, Canada
{thi.chau.giang.tran,thang.ledinh,francois.labelle,
rahma.chouchane}@uqtr.ca

[2] Danang University of Economics – The University of Danang, Danang, Vietnam

Abstract. Transitioning to a sustainable future demands a fundamental shift in resource management, emphasizing the principles of the circular economy (CE). This transformation is guided by the 10Rs framework, which includes strategies such as smarter product use and manufacture (Refuse, Rethink, Reduce), extending product lifespan (Reuse, Repair, Refurbish, Remanufacture, Repurpose), and sustainable material applications (Recycle, Recover). Intelligent systems (IS), encompassing adaptive applications and services capable of learning and evolving, present a transformative opportunity to optimize resource management and accelerate the transition to a CE. This systematic literature review explores the pivotal role of IS in enhancing resource efficiency and advancing sustainability. By analyzing current research, the study uncovers the synergies between these emerging fields and offers novel insights for fostering a circular and sustainable future.

Keywords: Intelligent System · Circular Economy · Sustainable Development

1 Introduction

The CE is a concept whose application comprises cutting back on resource consumption, eco-designing products so they can be readily disassembled and reused after use, extending product life, incorporating recyclables into products, and extracting resources from the waste flow [1]. Beside the undeniable advantages of the advantages from changing to CE, the linear "take, make, waste" model, where resources are extracted, manufactured into products, and ultimately discarded, is driving a global environmental crisis [2]. To address this crisis, this study examines the fundamental shift in resource management towards CE. The 10Rs framework provides a comprehensive set of principles for achieving circularity, encompassing a range of actions to minimize waste and maximize resource utilization [3]. This framework provides a roadmap for transitioning from a linear "take, make, waste" model to a CE that prioritizes resource efficiency and fosters a more sustainable future [4].

ISs, powered by artificial intelligence (AI), machine learning (ML), Internet of Things (IoT) and blockchain, are considered essential enablers of the CE [5]. ISs and

C. Yuan (Ed.): ICRER 2024, ESE, pp. 104–116, 2025.
https://doi.org/10.1007/978-3-031-93295-3_9

their ability to learn, adapt, and optimize processes through data analysis and automation [6], are poised to play a transformative role in the CE. They can provide the insights and tools needed to optimize resource consumption, enhance resource recovery, and promote product longevity [6].

However, since both CE and ISs are emerging fields, there is still a little focus on the systematic guidance on how ISs can be applied to strengthen the full potential of circular strategies. This study aims to bridge this gap by analyzing existing research to develop a comprehensive framework for the use of ISs to strengthen CE to provide valuable insights for researchers, practitioners, and policymakers seeking to harness the power of ISs to drive the transition towards a truly CE.

The paper begins with an introduction and background, followed by a methodology detailing the systematic review process. Findings cover applications of ISs in CE, with discussions and recommendations. It concludes with future research directions.

2 Background

This section presents the theoretical background for this study by introducing the 10Rs framework evolving from the initial "3R" approach (Reduce, Reuse, Recycle) [7] to address environmental concerns in manufacturing, which is a comprehensive set of principles for achieving circularity, and then presenting the concept of ISs, highlighting their potential to revolutionize resource management and promote sustainability.

2.1 The 10Rs Framework for Circularity

Table 1 includes strategies for 10Rs framework which offers a comprehensive set of guidelines for attaining circularity. This 10Rs framework provides a comprehensive set of principles for achieving circularity [3], encompassing strategies for:

- *Principle 1 (P1) - Smarter product use and manufacture*: Refuse, Rethink, Reduce - encouraging mindful consumption and efficient design.
- *Principle 2 (P2) - Extending the lifespan of products and their parts*: Reuse, Repair, Refurbish, Remanufacture, Repurpose - promoting durability, maintenance, and reintegration of components.
- *Principle 3 (P3) - Useful application of materials*: Recycle, Recover - maximizing resource recovery and minimizing waste generation.

Table 1. The base of circularity index [8].

R-strategy	Definition
Refuse	Making the decision not to use or buy needless or unsustainable things
Rethink	Putting traditional consumption habits to the test and creating long-lasting, reusable, and recyclable items
Reduce	Using as few materials and resources as possible over the course of a product's lifespan
Reuse	Giving objects a second lease of life by repurposing them for things like selling or charity

(continued)

Table 1. (*continued*)

R-strategy	Definition
Repair	Adding parts or fixing malfunctioning items to make them last longer
Refurbish	Covering the process of cleaning, maintaining, and replacing worn-out parts to return things to a like-new state
Remanufacturing	Involving the process of disassembling goods and reassembling them according to original design, frequently utilizing recycled materials
Repurpose	Putting items to new uses when they are no longer appropriate for their intended usage
Recycle	Creating new goods from recycled materials to cut down on the consumption of virgin resources
Recover	Recovering useful resources for recycling or reuse from waste streams

2.2 The 10Rs Framework for Circularity

By integrating recent technologies, including IoT, machine learning, deep learning, data mining, big data analytics, blockchain, and cloud computing, ISs are emerging as powerful tools for driving the CE [9]. In our study, the term of ISs is preferred to represent the ISs and their relevant applications and services. An interdisciplinary focus group has been created to propose and discuss about different types of ISs, which are shown in Table 2. These systems leverage data analysis, pattern recognition, and predictive modelling to optimize resource management, enhance resource recovery, and promote product longevity.

Table 2. Definition of ISs using in this study

Intelligent System	Definition
Recommender system (RS)	Leveraging user data and preferences to suggest relevant products, services, or information [10]
Decision support system (DSS)	Providing data-driven insights and support to decision-makers, helping them make informed choices regarding resource allocation, product design, and waste management [11–14]
Product-service system (PSS)	Shifting the focus from product ownership to product use, offering services alongside products to extend their lifespan and maximize their value [15–17]
Information management system (IMS)	Collecting, storing, analyzing, and disseminate data related to materials, products, and processes within CE initiatives [8, 18, 19]
Smart system (SS)	Combining multiple intelligent technologies and/or data-driven services to create integrated solutions that address complex challenges within the CE [15, 20]

3 Methodology

To assess the potential of ISs in driving the CE, this study undertakes a systematic literature review (SLR) [21] with focusing on the 10Rs strategies, a comprehensive framework encompassing the key principles of the CE, to understand how ISs can be applied to achieve circularity. Figure 1 presents the list of the steps in the SLR process and their results.

3.1 Research Objectives

Since the nature of CE is an interdisciplinary approach for encompassing various fields and different complex systems, the principles of service science have been adopted in this study [15]. Service science, or more precisely, service science, management, and engineering, is a term introduced to describe an interdisciplinary approach to the study and innovation of complex systems [22].

The primary goals of this research are founded on the various aspects of service science [15], with more focus on the Management and Science aspect, as follows:

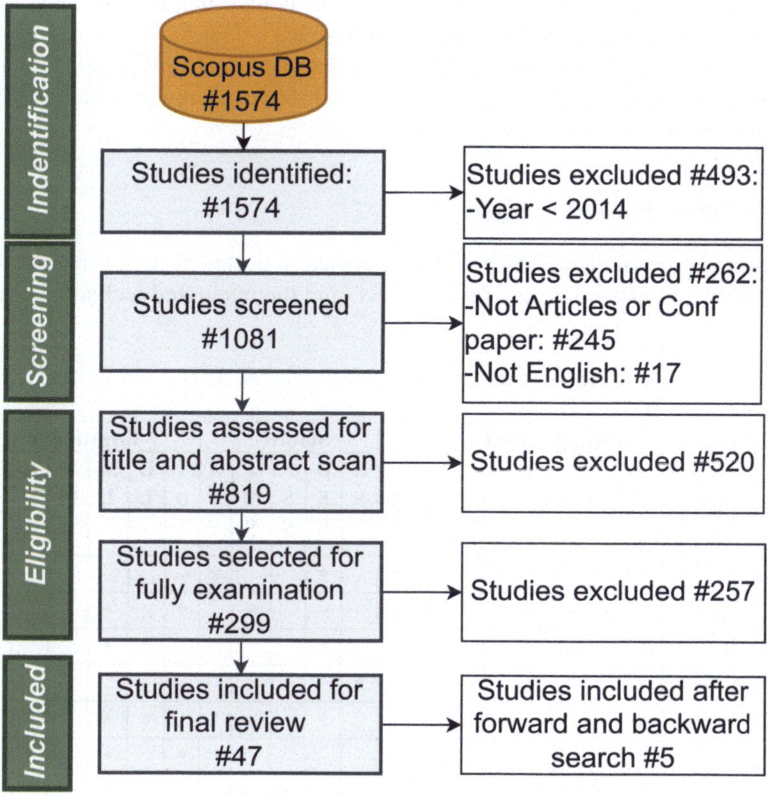

Fig. 1. Systematic literature review process

- *Management aspect*: examining the 10Rs framework and its three core principles, highlighting the role of ISs in achieving these goals.
- *Science aspect*: examining the underlying mechanisms of different types of ISs, applications, and tools, including RS, DSS, PSS, IMS, and smart systems.
- *Engineering aspect*: analyzing the implementation of machine learning, deep learning, IoT, data mining, big data analytics, blockchain, and cloud computing in the ISs.

3.2 Search Strategy

The Scopus database was used to compile the research studies for this investigation (Fig. 1). This database was selected with consideration for its breadth and academic value in mind.

To ensure a comprehensive and systematic review, a three-part keyword strategy was used to identify relevant articles by searching for specific terms in their titles, keywords, or abstracts. The first group of keywords focused on the core theme of the review: CE, including CE, circular business, sustainable economy, closed-loop economy, sustainable development, CE principles, and 10R strategies; The second group aimed at kinds of IS, including RS, DSS, PSS, IMS, CPS, and smart system. The third group are keywords related to technologies: blockchain, cloud computing, IoT, machine learning, deep learning, data mining, and big data analytics.

Figure 1 describes the process of SLR for papers selection in this study. After reviewing the titles and abstracts, 47 publications were included in the review following full-text quality evaluation.

3.3 Data Extraction and Analysis

Table 3 provides a visualization about the intersection of the 10 Rs framework to solve in CE with the corresponding used type of ISs (and the integrated technologies) across

Table 3. Publications included in the study

Article	Management										Science					Engineering						
	P1		P2						P3		RS	DSS	PSS	IMS	SS	IoT	DL	ML	DM	BDA	BC	CC
	R0	R1	R2	R3	R4	R5	R6	R7	R8	R9												
P1 [23]		*	*	*		*	*			*		*			*				*			
P2 [8]	*	*	*	*	*	*	*	*	*	*				*		*		*				
P3 [24]			*	*						*		*					*	*				
P4 [25]			*							*	*								*			
P5 [26]			*						*			*		*			*	*		*		
P6 [27]		*	*	*					*	*					*	*		*		*		*
P7 [28]		*										*						*	*			
P8 [29]		*	*	*						*		*					*		*			
P9 [30]		*	*						*	*		*						*	*	*		
P10 [31]			*						*	*					*		*	*				
P11 [32]		*	*						*	*					*		*	*	*			
P12 [33]		*	*						*	*				*				*	*		*	
P13 [34]		*	*	*	*	*	*			*		*		*		*					*	
P14 [35]		*	*	*			*			*					*						*	
P15 [18]		*	*	*					*	*				*		*					*	
P16 [36]			*	*						*		*									*	
P17 [37]			*						*	*		*						*	*			
P18 [10]			*	*						*	*				*				*			
P19 [38]		*	*							*					*		*	*				*

(continued)

Table 3. (*continued*)

Article	Management										Science					Engineering						
	P1			P2					P3		RS	DSS	PSS	IMS	SS	IoT	DL	ML	DM	BDA	BCC	CC
	R0	R1	R2	R3	R4	R5	R6	R7	R8	R9												
P20 [11]			*	*			*			*		*		*				*				
P21 [39]				*						*				*				*	*	*		
P22 [40]		*	*	*					*	*				*				*				
P23 [19]			*						*	*				*	*			*		*		
P24 [41]		*	*	*						*		*		*					*	*		
P25 [42]		*	*	*		*				*				*		*		*	*	*		
P26 [15]			*	*	*	*	*			*			*							*		
P27 [43]			*	*	*	*	*			*				*							*	
P28 [44]		*	*	*					*	*		*				*						
P29 [45]		*	*	*		*	*			*				*							*	
P30 [46]		*	*	*	*		*		*	*				*				*	*			
P31 [47]		*	*	*					*	*		*				*			*			
P32 [48]		*	*						*									*	*			
P33 [49]			*	*					*	*		*						*	*	*		
P34 [50]			*	*						*		*		*				*	*			
P35 [51]				*						*		*			*			*	*			
P36 [16]				*	*	*	*			*			*									
P37 [52]		*	*	*	*										*			*	*			
P38 [53]			*	*	*	*	*	*	*	*					*	*					*	*
P39 [54]			*	*			*		*	*					*			*				
P40 [17]		*	*	*									*						*			
P41 [12]				*	*	*	*			*		*				*	*		*			
P42 [13]			*							*		*						*	*			
P43 [55]			*							*		*						*	*			
P44 [56]		*	*	*	*		*			*				*				*	*			
P45 [14]		*	*	*	*		*			*		*				*		*				
P46 [57]		*	*						*	*					*			*	*			
P47 [58]	*	*	*	*						*		*										
Total: 47	**2**	**26**	**42**	**33**	**11**	**10**	**15**	**2**	**19**	**42**	**2**	**21**	**3**	**16**	**13**	**10**	**7**	**28**	**25**	**9**	**8**	**3**

47 research articles. It also summarizes the AI/ML technologies and different types of ISs investigated in the reviewed studies.

4 Findings

4.1 Application Domains of Intelligent Systems

- *R0 – Refuse*: DSSs and IMSs present a promising solution by automating sorting recyclable materials from refusing, utilizing computer vision, machine learning, optical recognition, and intelligent evaluation algorithms to identify and separate various waste fractions with high accuracy, highlighting the potential of DSSs and IMSs to significantly enhance the efficiency of refuse management [8, 58].
- *R1 – Rethink*: Smart systems utilizing machine learning and deep learning, to move away from traditional, manual sorting methods to "rethink" waste management management [32, 38]. Besides, the importance of data-driven DSS is emphasized when using data analysis, modeling, and prediction to challenge traditional approaches in sectors like construction and waste management in terms of Rethink [28–30]. Blockchain in IMSs also "rethink" the CE by fostering transparency and traceability, creating secure and tamper-proof records of material lifecycles [18, 33].
- *R2 – Reduce*: A RS [10] connects businesses with complementary needs, turning waste into valuable inputs. DSSs [11, 19] help select sustainable suppliers committed to CE practices. Smart systems [27, 33, 38, 57] leverage AI, blockchain, and IoT to optimize waste management processes, reducing the demand for virgin materials and waste disposal.
- *R3 – Reuse*: A RS [10] identifies potential reuse opportunities for waste materials, while a DSS [11] helps businesses select sustainable suppliers committed to CE practices, including reuse. An AI-powered DSS [12] guides businesses in transitioning to circular supply chains, classifying products and components for optimal reuse. Blockchain-based platform [43] improves the quality of recycled plastics, making them more viable for reuse in manufacturing.
- *R4 – Repair*: Smart systems are playing an increasingly important role in enhancing repair processes and promoting a CE. An AI-powered damage detection system [52] identifies repair needs early, extending product lifespans. Research [16] highlights the importance of "Repair, Upgrade, and Debugging Services" (RUD) for extending product lifecycles, suggesting the potential for smart monitoring systems. IMS [34], leveraging blockchain and IoT, collects and manages repair data, enabling secure and transparent tracking of repair history, optimizing spare parts management, and facilitating repair networks.
- *R5 – Refurbish*: Expert system [23] and PSS [15] can guide refurbishment processes, facilitating strategic planning and detailing specific refurbishment services. Blockchain creates a secure and transparent record of a product's refurbishment history, increasing consumer trust and streamlining processes. A "Refurbishment Certification Tool" [43], built on a blockchain-based PLM system, generates trustworthy certificates for refurbished products, increasing consumer confidence in their quality.
- *R6 – Remanufacture*: DSS [12] can optimize remanufacturing processes by optimizing logistics, identifying candidates, and optimizing operations. Expert system [23] can guide remanufacturing strategies by analyzing product lifecycles and material flows, identifying opportunities and designing strategies. PSSs [15, 16] can integrate remanufacturing by defining "Service Structures" and offering remanufacturing services as part of their offerings.

- *R7 – Repurpose*: Smart systems, built on blockchain and IoT, can enhance repurposing by tracking product condition and history, optimizing logistics for efficient routing, and enhancing transparency to build trust in repurposed products [53].
- *R8 – Recycle*: Smart systems, including AI-powered robotic sorting [31] and deep learning for material classification [32], are enhancing recycling operations. MS leveraging blockchain and multi-sensor data fusion [33, 18], increase transparency and traceability, improve efficiency, and incentivize recycling. DSS using machine learning [37], predicts waste generation, optimizing processes and reducing landfill waste. Other systems and applications relevant to recycling are being developed for food waste recycling, automated waste segregation, and plastic CE frameworks.
- *R9 – Recover*: A RS [10] facilitates industrial symbiosis by matching resource needs and offers, turning waste into valuable resources. DSSs [11, 58] aid in informed waste management decisions, optimizing resource allocation, and promoting recovery. PSS [15] promotes circularity by design and extended lifecycles. IMSs [10, 54] are crucial for data-driven sustainability, enabling material flow tracking, performance monitoring, and decision-making.

4.2 Application Domains of Intelligent Systems

- *Recommender systems*: RSs are emerging as tools for promoting CE principles. A RS [25] helps reduce household food waste. These RSs leverage data analysis and learning to refine recommendations, demonstrating the potential of RSs to drive CE practices across various sectors.
- *Decision support systems*: DSSs powered by machine learning can be used for sustainable supplier selection and optimizing waste treatment and transportation supply chains [11]. CEBMES [23] is an expert system that guides manufacturing companies towards circular business models. A group DSS for transitioning to a CE model in solid waste management [58] integrates problem structuring methods, MCDM, and game theory to facilitate collaborative decision-making.
- *Product service systems*: PSSs are increasingly recognized as key enablers of sustainability and CE principles. A study emphasizes the importance of identifying and prioritizing customer value requirements when designing a PSS for circularity, particularly in the wind energy sector, using Fuzzy AHP to rank the importance of different value requirements [16].
- *Information management systems*: IMSs are crucial for supporting waste management and CE initiatives. AI and blockchain tools are being developed to provide manufacturers with reliable information about recycled feedstock [33], while blockchain technology is being explored to track professional electrical and electronic equipment throughout its lifecycle [34]. Smart information systems are essential for implementing the CE [8], capturing data related to resource preservation, material flows, and environmental impacts.
- *Smart systems*: Smart systems are being developed to optimize waste management and promote CE principles. AI-based expert systems like CEBMES [23] guide companies in developing circular business models, while ML models are used to predict and optimize various aspects of waste management [38, 51]. In supply chain optimization, the concept of a virtual closed-loop supply chain is introduced [53], where digital technologies augment physical operations.

5 Discussion

5.1 The Potential of ISs for Circularity

The potential of ISs for driving CE principles is undeniable. These systems, leveraging a powerful toolkit of AI, ML, blockchain, and IoT, are transforming various aspects of the CE lifecycle. From revolutionizing waste management by automating sorting processes [27], utilizing computer vision and machine learning [43, 49], and optimizing resource allocation [19, 41, 42], to extending product lifespans through reuse, repair, and remanufacturing [29–31, 37, 49, 53], these technologies are demonstrating their ability to create a more sustainable future.

Furthermore, ISs are optimizing resource utilization through expert systems [23] and PSSs [12] that guide refurbishment processes, and IMS [45] that leverages blockchain and IoT to enhance data management throughout a product's lifecycle.

5.2 Challenges and Opportunities

The integration of ISs into the CE presents both significant challenges and exciting opportunities. Ensuring high-quality, accessible data, addressing security and privacy concerns, and establishing clear policy frameworks are crucial for realizing the full potential of these technologies [36, 42, 44, 46].

On the other side, ISs offer significant opportunities for enhanced efficiency and optimization in waste management, resource utilization, and product lifecycle management, leading to cost savings and reduced environmental impact [19, 27, 30, 37, 41, 45, 48]. By addressing the challenges and capitalizing on the opportunities, we can harness the power of ISs to create a more sustainable and equitable future.

5.3 Research Gap and Future Work

While the findings highlight the significant potential of ISs for circularity, several research gaps and areas for future work remain. Further research is needed to develop robust frameworks for integrating ISs into existing circular systems [34]. More research is also needed to address ethical considerations, ensure equitable access to these technologies, and develop robust frameworks for integrating them into existing systems [42, 46]. Finally, research is needed to develop methods for evaluating the effectiveness of ISs in driving circularity and measuring their impact on environmental and social outcomes [10, 11, 31].

6 Conclusion

The paper explores the significant potential of ISs, including RS, DSS, PSS, IMS, and SS, to enhance the implementation of the 10Rs framework in achieving CE principles. Key applications of these ISs are identified across various stages of the 10Rs framework, demonstrating their ability to optimize resource utilization, reduce waste, extend product lifespans, and drive innovation in circular business models [4, 51, 52]. The study's analysis highlights the potential of ISs to revolutionize waste management through automated sorting and resource allocation, extend product lifespans through reuse, repair, and remanufacturing, and optimize resource utilization through expert systems, PSS, and IMS.

However, it is acknowledged that challenges remain, including data quality, security, policy frameworks, and equitable access to these technologies [11, 17, 27, 37, 47, 53]. Addressing these challenges requires a collaborative effort among researchers, policymakers, and industry stakeholders. This research offers valuable insights for businesses seeking to implement CE strategies by providing practical applications of ISs to improve efficiency, reduce waste, and enhance resource management.

Funding. This research was supported in part by RDÉE Canada (Réseau de Développement Économique et d'Employabilité), via the Enabling Fund for Official Language Minority Communities, and by the RRECQ (Réseau de Recherche en Économie Circulaire du Québec) via the FRQ (Fonds de Recherche du Québec).

References

1. Chennak, A., Giannakas, K., Awada, T.: On the economics of the transition to a circular economy. Circular Economy Sustainability 1–17 (2023)
2. Elisha, O.D.: Moving beyond take-make-dispose to take-make-use for sustainable economy. Int. J. Sci. Res. Educ. **13**, 497–516 (2020)
3. Potting, J., Hekkert, M.P., Worrell, E., Hanemaaijer, A.: Circular economy: measuring innovation in the product chain. Planbureau voor de Leefomgeving (2017)
4. Brändström, J., Jazairy, A., Roos Lindgreen, E.: Barriers to adopting circular business models: a cross-sectoral analysis. Bus. Strategy Environ. (2024)
5. Sterev, N., Biolcheva, P.: Intelligent systems in the circular economy. In: Proceedings of the VIth International Conference on Management and Strategic Decisions (ICGSM), pp. 8–10. (2023)
6. Tzafestas, S., Borne, P.: Intelligent systems, control and automation: science and engineering. Springer (2012)
7. Union, E.: Directive 2008/98/EC of the European Parliament and of the Council of 19 November 2008 on waste and repealing certain directives. Off. J. Eur. Union **51**, 3–30 (2008)
8. Reich, R.H., Alaerts, L., Van Acker, K.: The circular economy as a requirement for smart information system designs. In: 2023 19th International Conference on Distributed Computing in Smart Systems and the Internet of Things (DCOSS-IoT), pp. 359–366. IEEE, (2023)
9. Rossi, J., Bianchini, A., Guarnieri, P.: Circular economy model enhanced by intelligent assets from industry 4.0: The proposition of an innovative tool to analyze case studies. Sustainability **12**, 7147 (20t0)
10. Gatzioura, A., Sànchez-Marrè, M., Gibert, K.: A hybrid recommender system to improve circular economy in industrial symbiotic networks. Energies **12**, 3546 (2019)
11. Alavi, B., Tavana, M., Mina, H.: A dynamic decision support system for sustainable supplier selection in circular economy. Sustain. Prod. Consum. **27**, 905–920 (2021)
12. Mboli, J.S., Thakker, D., Mishra, J.L.: Artificial intelligence-powered decisions support system for circular economy business models. In: ICEIS, no. 1, pp. 656–666 (2023)
13. Agrawal, N., Rabiee, M., Jabbari, M.: Contextual relationships in Juran's quality principles for business sustainable growth under circular economy perspective: a decision support system approach. Ann. Oper. Res. 1–31 (2023)

14. Mboli, J.S., Thakker, D., Mishra, J.L.: An Internet of Things-enabled decision support system for circular economy business model. Softw. Pract. Exp. **52**, 772–787 (2022)
15. Halstenberg, F.A., Lindow, K., Stark, R.: Leveraging circular economy through a methodology for smart service systems engineering. Sustainability **11**, 3517 (2019)
16. Nag, U., Sharma, S.K., Padhi, S.S.: Evaluating value requirement for industrial product-service system in circular economy for wind power-based renewable energy firms. J. Clean. Prod. **340**, 130689 (2022)
17. Metic, J., Guzzo, D., Kopainsky, B., McAloone, T.C., Pigosso, D.C.: A simulation-based approach for investigating the dynamics of rebound effects in the circular economy: a case of use-oriented product/service system. J. Environ. Manag. **365**, 121627 (2024)
18. Castiglione, A., Cimmino, L., Di Nardo, M., Murino, T.: A framework for achieving a circular economy using the blockchain technology in a sustainable waste management system. Comput. Ind. Eng. **180**, 109263 (2023)
19. Lanzalonga, F., Marseglia, R., Irace, A., Biancone, P.P.: The application of artificial intelligence in waste management: understanding the potential of data-driven approaches for the circular economy paradigm. Manag. Decis. (2024)
20. Yamarthi, N.R., et al.: A novel IoT-based smart and security system model for large scale farm sustainability. J. Adv. Inf. Technol. **15** (2024)
21. Levy, Y., Ellis, T.J.: A systems approach to conduct an effective literature review in support of information systems research. Inf. Sci. **9** (2006)
22. Spohrer, J., Maglio, P.P., Bailey, J., Gruhl, D.: Steps toward a science of service systems. Computer **40**, 71–77 (2007)
23. Pieroni, M.P., McAloone, T.C., Borgianni, Y., Maccioni, L., Pigosso, D.C.: An expert system for circular economy business modelling: advising manufacturing companies in decoupling value creation from resource consumption. Sustain. Prod. Consum. **27**, 534–550 (2021)
24. Yu, M., Zheng, J.: Introducing artificial/computational intelligence-derived non-parametric transfer functions for the implementation of dynamic circular economy decision-making systems. Electronics **12**, 1050 (2023)
25. Takahashi, O., Sasaki, S.: A dinner menu & recipe recommendation system with food expiration-date notification functions. In: 2022 13th International Congress on Advanced Applied Informatics Winter (IIAI-AAI-Winter), pp. 230–235. IEEE (2022)
26. Matheri, A.N., Ntuli, F., Ngila, J.C., Seodigeng, T., Zvinowanda, C.: Performance prediction of trace metals and cod in wastewater treatment using artificial neural network. Comput. Chem. Eng. **149**, 107308 (2021)
27. Chen, X.: Machine learning approach for a circular economy with waste recycling in smart cities. Energy Rep. **8**, 3127–3140 (2022)
28. Prioux, N., Ouaret, R., Hetreux, G., Belaud, J.-P.: Environmental assessment coupled with machine learning for circular economy. Clean Technol. Environ. Policy **25**, 689–702 (2023)
29. Akanbi, L.A., Oyedele, A.O., Oyedele, L.O., Salami, R.O.: Deep learning model for demolition waste prediction in a circular economy. J. Clean. Prod. **274**, 122843 (2020)
30. Chin, H.H., Varbanov, P.S., You, F., Sher, F., Klemeš, J.J.: Plastic circular economy framework using hybrid machine learning and pinch analysis. Resour. Conserv. Recycl. **184**, 106387 (2022)
31. Wilts, H., Garcia, B.R., Garlito, R.G., Gómez, L.S., Prieto, E.G.: Artificial intelligence in the sorting of municipal waste as an enabler of the circular economy. Resources **10**, 28 (2021)
32. Nañez Alonso, S.L., Reier Forradellas, R.F., Pi Morell, O., Jorge-Vázquez, J.: Digitalization, circular economy and environmental sustainability: the application of Artificial Intelligence in the efficient self-management of waste. Sustainability **13**, 2092 (2021)
33. Chidepatil, A., Bindra, P., Kulkarni, D., Qazi, M., Kshirsagar, M., Sankaran, K.: From trash to cash: how blockchain and multi-sensor-driven artificial intelligence can transform circular economy of plastic waste? Adm. Sci. **10**, 23 (2020)

34. Magrini, C., et al.: Using internet of things and distributed ledger technology for digital circular economy enablement: the case of electronic equipment. Sustainability **13**, 4982 (2021)
35. Shojaei, A., Ketabi, R., Razkenari, M., Hakim, H., Wang, J.: Enabling a circular economy in the built environment sector through blockchain technology. J. Clean. Prod. **294**, 126352 (2021)
36. Basile, D., D'Adamo, I., Goretti, V., Rosa, P.: Digitalizing circular economy through blockchains: the blockchain circular economy index. J. Ind. Prod. Eng. **40**, 233–245 (2023)
37. Cha, G.-W., Park, C.-W., Kim, Y.-C.: Optimal machine learning model to predict demolition waste generation for a circular economy. Sustainability **16**, 7064 (2024)
38. Mohammed, M.A., et al.: Automated waste-sorting and recycling classification using artificial neural network and features fusion: a digital-enabled circular economy vision for smart cities. Multimed. Tools Appl. **82**, 39617–39632 (2023)
39. Xue, Y., Kambhampati, C., Cheng, Y., Mishra, N., Wulandhari, N., Deutz, P.: A LDA-based social media data mining framework for plastic circular economy. Int. J. Comput. Intell. Syst. **17**, 8 (2024)
40. Hala, H., Anass, C., Youssef, B.: Machine learning for the future integration of the circular economy in waste transportation and treatment supply chain. IFAC-PapersOnLine **55**, 49–54 (2022)
41. Pamučar, D., Durán-Romero, G., Yazdani, M., López, A.M.: A decision analysis model for smart mobility system development under circular economy approach. Socioecon. Plann. Sci. **86**, 101474 (2023)
42. Munir, M.T., Li, B., Naqvi, M., Nizami, A.-S.: Green loops and clean skies: optimizing municipal solid waste management using data science for a circular economy. Environ. Res. 117786 (2023)
43. Regueiro, C., Gómez-Goiri, A., Pedrosa, N., Semertzidis, C., Iturbe, E., Mansell, J.: Blockchain-based refurbishment certification system for enhancing the circular economy. Blockchain: Res. Appl. **5**, 100172 (2024)
44. Iqbal, M.W., Kang, Y.: Circular economy of food: a secondary supply chain model on food waste management incorporating IoT based technology. J. Clean. Prod. **435**, 140566 (2024)
45. Cerchione, R.: Design and evaluation of a blockchain-based system for increasing customer engagement in circular economy. Corporate Social Responsibility and Environmental Management (2024)
46. Platon, V., et al.: New evidence about artificial intelligence and eco-investment as boosters of the circular economy. Environ. Technol. Innov. 103685 (2024)
47. Govindan, K., Asgari, F., Fard, F.S.N., Mina, H.: Application of IoT technology for enhancing the consumer willingness to return E-waste for achieving circular economy: a Lagrangian relaxation approach. J. Clean. Prod. **459**, 142421 (2024)
48. Ashraf, W.M., et al.: Machine learning assisted improved desalination pilot system design and experimentation for the circular economy. J. Water Process Eng. **63**, 105535 (2024)
49. Liu, T., Guan, X., Wang, Z., Qin, T., Sun, R., Wang, Y.: Optimizing green supply chain circular economy in smart cities with integrated machine learning technology. Heliyon **10** (2024)
50. Chen, M., Liu, Q., Huang, S., Dang, C.: Environmental cost control system of manufacturing enterprises using artificial intelligence based on value chain of circular economy. Enterp. Inf. Syst. **16**, 1856422 (2022)
51. Dan, Z., Song, A., Yu, X., Zhou, Y.: Electrification-driven circular economy with machine learning-based multi-scale and cross-scale modelling approach. Energy **299**, 131469 (2024)
52. Gordan, M., Sabbagh-Yazdi, S.-R., Ghaedi, K., Ismail, Z.: A damage detection approach in the era of industry 4.0 using the relationship between circular economy, data mining, and artificial intelligence. Adv. Civ. Eng. **2023**, 3067824 (2023)

53. Prajapati, D., Jauhar, S.K., Gunasekaran, A., Kamble, S.S., Pratap, S.: Blockchain and IoT embedded sustainable virtual closed-loop supply chain in E-commerce towards the circular economy. Comput. Ind. Eng. **172**, 108530 (2022)
54. Oliveira Neto, G.C.D., de Araujo, S.A., Gomes, R.A., Alliprandini, D.H., Flausino, F.R., Amorim, M.: Simulation of electronic waste reverse chains for the sao paulo circular economy: an artificial intelligence-based approach for economic and environmental optimizations. Sensors **23**, 9046 (2023)
55. Zhao, J., Zhu, Q.: China's energy structure optimization based on data mining from the perspective of circular economy: take Qujing City as an example. Mob. Inf. Syst. **2022**, 7905576 (2022)
56. Walter, S., Braun, A., Louw, L.: Development of a product return process in the context of the circular economy with the help of machine learning. Procedia CIRP **120**, 1498–1503 (2023)
57. Kumar, N.M., et al.: Artificial intelligence-based solution for sorting COVID related medical waste streams and supporting data-driven decisions for smart circular economy practice. Process. Saf. Environ. Prot. **152**, 482–494 (2021)
58. Silva, W.D.O., Morais, D.C.: A group decision support system to share responsibilities towards a circular economy transition for solid waste management in developing countries. In: 2021 IEEE International Conference on Systems, Man, and Cybernetics (SMC), pp. 2221–2226. IEEE (2021)

Environmental Engineering Technologies and Applications

A New Method for Determining the Stability Velocity of Gas-Oil Interface on Gas Assisted Gravity Drainage Considered Integrate Effect of Engineering and Geology

Wenshuang Geng(✉), Lixin Yuan, and Xiaoli Ma

Exploration and Development Research Institute of PetroChina Jidong Oilfield Company, Tangshan 063004, Hebei, China
gengwenshuang@126.com

Abstract. Dietz model is a common method to determine whether the gas-liquid interface is stable in inclined reservoirs, but this model does not consider the influence of elevation difference between injection and production wells, well spacing, reservoir rhythm and pressure drop funnel of liquid production wells on the stability of gas-liquid interface. The maximum angle of gas-liquid interface is quantitatively expressed by introducing the elevation difference between injection and production wells and the plane well spacing. According to the rhythm characteristics and injection-production ratio, a new method for determining the stability velocity of gravity-assisted flooding of gas injection is further given, and the stability conditions of secondary gas top of LBIV oil group are analyzed. The results show that this method can be used for the dynamic analysis of enhanced oil recovery by gas injection gravity-assisted flooding after high-angle water flooding.

Keywords: Gas injection · Gas-liquid interface · Gravity-assisted drive · Enhanced Oil Recovery · High angle reservoir

1 Review on the Determination of Reasonable Velocity of Gas Injection Gravity Stable Flooding

The formation of stable secondary gas roof and the stable interface of gas flooding front edge are one of the keys to enhanced oil recovery by gas injection after high-angle water flooding. The stability of gas flooding front is mostly studied in miscible flooding, and the analytical theoretical methods include inclined reservoir model (Dietz model) [1] and vertical downward miscible model (Dumore) [2]. Dumore method is a recommended method in Chinas industry standards. Both of them are based on the basic theory of oil-gas phase separation flow, and consider that the gas tongue enters the oil area beyond the crude oil intrusion at the interface, so that their respective expression models of critical gas injection velocity with stable gas-liquid interface are obtained. Relatively speaking, apart from the influence of reservoir dip angle, the difference between the two critical gas injection velocities is only the phase permeability used by the former when considering

C. Yuan (Ed.): ICRER 2024, ESE, pp. 119–131, 2025.
https://doi.org/10.1007/978-3-031-93295-3_10

the oil flow velocity, while the absolute permeability used by the latter, that is, under the same conditions, the critical velocity of Dietz model is smaller than that of Dumore model. But in fact, the stability of the gas-liquid interface cannot be achieved below the critical velocity. If the mass transfer at the gas-liquid interface is considered, R.L. Perrine [3] used the gas-liquid interface disturbance method in 1961 to obtain the completely stable condition; Dumore gives the same completely stable characteristic velocity by considering the increase of the transition zone with depth, and gives the completely stable characteristic velocity by a large number of experimental statistics. The method of deriving the pressure gradient of the transition zone to the depth is considered, and the completely stable characteristic velocity is obtained by a large number of experiments and statisticsRatio of steady velocity to critical velocity [2].

But in fact, how to accurately determine the critical speed has always been a difficult problem. In 1962, J. W. Sheldon & F. J. Based on the Taylor expansion of the gas-liquid interface migration along the downward direction of the reservoir in the Dietz geometric model, the two men gave the expressions of the downward velocity and stability conditions of the interface, but the critical velocity had no trigonometric function terms of gravity acceleration and inclination angle, and the trigonometric function was directly replaced by the inclination angle [4]. E. L. The critical velocity given by Dougherty (1963) is the molecular subtraction of the fraction of the gas phase entering the oil phase, and the viscosity of the gas phase is divided by an empirical correction constant to consider the influence of reservoir heterogeneous dispersion [5]. R.L. When Cardenas et al. designed the reasonable injection rate of CO2 injection for the nearshore inclined reservoir in Louisiana State, they divided the Dietz critical velocity by the reservoir porosity, and considered that the critical velocity obtained by this calculation was 33% higher than the stable velocity [6].

J.G.J. Ypma (1984) holds that when considering reservoir heterogeneity, the gas-liquid flow will have a certain angle with the vertical direction, which is directly related to the ratio of permeability along the reservoir direction to the permeability in the reservoir normal direction and the dip angle of the reservoir. The maximum velocity of gravity stable flooding should be equal to the critical velocity divided by the effective flow porosity and the cosine of this angle [7]. The key of this method lies in the determination of bedding horizontal permeability and vertical permeability of layered reservoirs. F.J. Fayers (1990) gave an extended Dietz model in the form of partial differential equation of gas-liquid interface downward velocity, which can consider the migration and evolution of gas-liquid interface to a certain extent when the vertical permeability and horizontal permeability ratios of reservoirs vary. However, as the permeability ratio decreases below about 0.3, the model gradually fails [8], and this method needs to be solved by numerical solution. In 2002, Xiong Yu et al. considered the method of determining the reasonable gas injection rate of gas-flooded multi-layer reservoirs after water flooding. The main difference is that when the gas-liquid interface moves, the fluidity of oil and gas on both sides of the interface is considered respectively. The oil phase fluidity under the residual oil of water flooding and the gas phase fluidity under the residual liquid phase of gas flooding. It is considered that the high permeability thin layer in multilayer is the key to control the reasonable gas injection rate [9]. In 2011, N.A. F. Although El-Khatib also considers the two-phase displacement interface under inclined multilayer heterogeneity,but it is aimed at interface movement under water flooding conditions [10].

In 2018, Ren Shaoran et al. determined the gravity stability conditions through the critical velocity deformation of Dietz model and experimental methods [11]; The specific method is to ignore the gas viscosity and the oil permeability, take the absolute permeability, and take the dip angle of gas-liquid interface equal to the dip angle of reservoir as the limit condition, so as to give the stable flooding velocity condition. But in fact, because absolute permeability replaces oil phase permeability, the ratio of permeability to oil phase viscosity will be greatly increased, and the effect of ignoring gas phase viscosity can be completely ignored. Wang Rui (2021) based on the gas-liquid phase separation flow theory and considering the gas-liquid pressure difference at the gas-liquid interface, gave a calculation method for the critical velocity of the artificial gas cap at the injection-production ratio of 1: 1 [12]. The main difference is that the molecular density difference term is 2 times the difference between the oil phase density and the gas phase density, and the oil phase permeability is used instead of the absolute permeability, so the predicted critical velocity is slightly smaller than that of Dietz model.

Based on the above, it can be seen that the prediction method of reasonable gas injection rate has not made much progress. Until recently (2022), the early method based on the product of Dietz critical velocity and gas intake area was still used in the design of reasonable gas injection volume in actual reservoirs [13]. Moreover, these methods are based on the basic assumption that the injection-production ratio is 1, and do not consider the influence of elevation difference between injection-production wells, well spacing, pressure drop funnel of liquid production wells and different rhythm structures of reservoirs. It brings great difficulties to the dynamic analysis of gas injection-assisted gravity flooding and the regulation of gas-liquid interface in reservoirs after high-angle water flooding, such as Liubei IV oil group.

2 Establishment of Improved Dietz Model Considering Engineering Factors

2.1 Understanding Based on Profile Physical Simulation and Numerical Simulation

In order to study the influence of pressure drop funnel in oil production wells on the stability of gas-liquid interface between injection and production wells, two-dimensional numerical simulation and physical simulation were used respectively. The physical simulation of the profile adopts a sand-filled cementation two-dimensional model with a thickness of 2 cm and a vertical height of 20 cm. The porosity and permeability distribution of the cementation model are tested by drilling cores after testing. The vertical physical properties and average physical properties of the profile model are filled according to the parameters of Liubei IV oil group, and the test pressure and temperature are the current reservoir formation pressure and temperature, which are 30MPa and 95.6 °C respectively. During the test, it is shown (as shown in Fig. 1) that when the relatively stable gas-liquid interface advances to the edge of the pressure drop funnel close to the production well, unstable tongue entry phenomenon will occur.

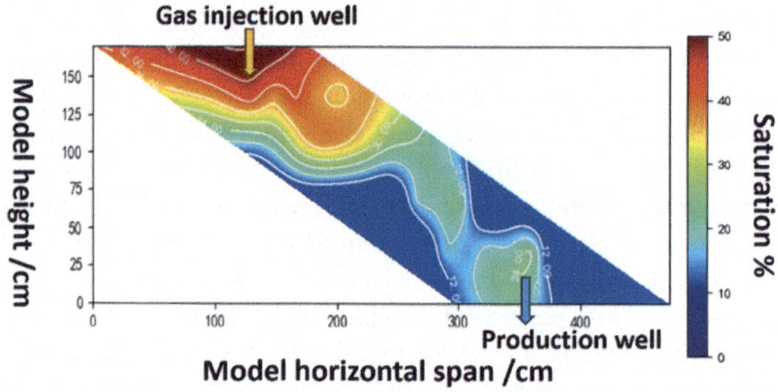

Fig. 1. Two-dimensional physical simulation of stability of artificial gas cap

In this study, a detailed two-dimensional numerical simulation is used to analyze the relationship between the stability of gas-liquid interface and the host ratio. When the injection rate is constant and the injection-production ratio is changed by reducing the production rate at the production end, it can be seen by looking at the stability of the gas-liquid interface (see Fig. 2) that the increase of the injection-production ratio is conducive to the generation of secondary gas roof in high-angle reservoirs and the formation of gravity-assisted flooding, that is, when the production well is produced at low speed, the gas channeling will be delayed due to the smaller pressure drop funnel.

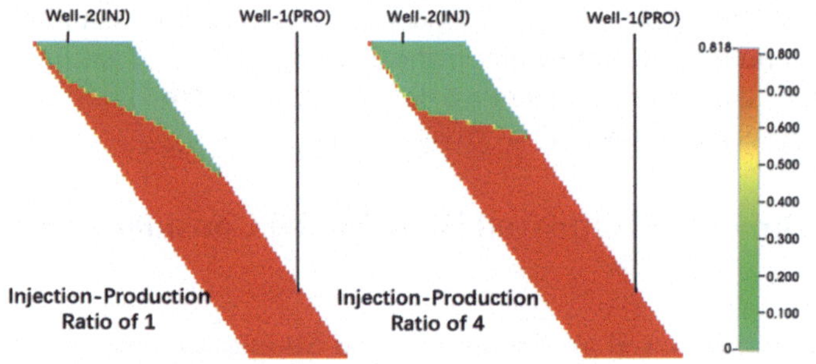

Fig. 2. Two-dimensional numerical simulation of gas-liquid interface stability during top gas injection

Using two-dimensional numerical simulation to simulate the stable data under different gas injection velocities and injection-production ratios, the stable area and critical line shown in Fig. 3 can be obtained without considering the similarity between the model and the actual formation, that is, with the increase of gas injection velocity, the injection-production ratio must also increase to maintain interface stability, which is often unrealistic for actual reservoirs, because not only the annual output of actual reservoirs has corresponding planning, but also the gas injection volume is limited by the design gas injection volume, pipeline network adaptation requirements and gas source allocation. The planned annual production requirements will inevitably lead to the fact that the oil production well must have a certain pressure drop interval and oil drainage radius (pressure drop funnel range).

From the results shown in Fig. 2, it can be known that when the elevation difference between injection and production wells decreases, the plane well spacing will decrease, which will also affect the reasonable gas injection rate. Therefore, for the actual gravity-assisted flooding of gas injection at the top of high-angle reservoirs, the reasonable gas injection velocity must be the function of elevation difference between injection and production wells, well spacing, liquid production and reservoir rhythm. At the same time, it is also a function of parameters such as high permeability thin layer and the ratio of vertical permeability to horizontal permeability in reservoir interval [9, 14], and it also has a reasonable oil recovery rate when keeping the reservoir liquid zone in a single-phase state [14].

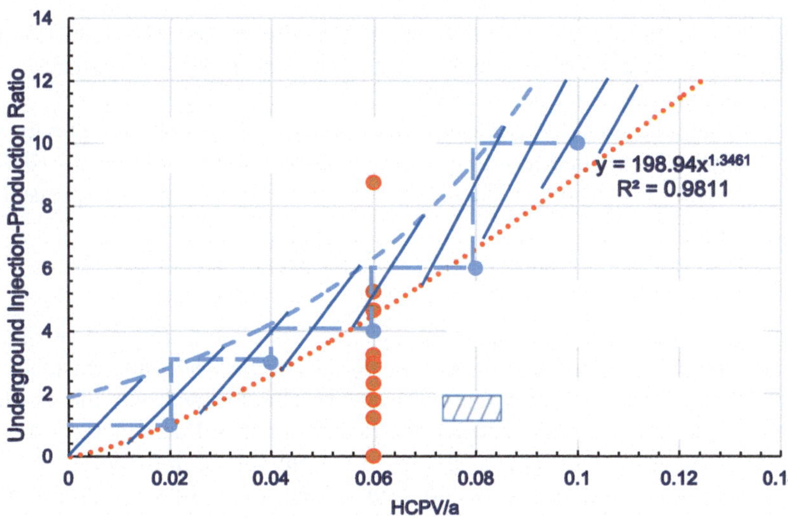

Fig. 3. Discrimination plate of gravity stable flooding of gas injection in high dip angle reservoir

2.2 Improved Dietz Model Considering Engineering Factors

From the previous Sect. 1, we know that Dietz model is the basic model for determining the reasonable gas injection rate of inclined or high-angle reservoirs, and most researchers basically improve this model [4 ~ 12]. The basic geometric relationship of the Dietz model can be simplified to the form shown in Fig. 4.

When considering the elevation difference (H) and well spacing (D) between injection and production wells, it is evident that the elevation difference and well spacing limit the maximum value of the included angle β:

$$\tan\beta_{\max} = \mathrm{H/D} \tag{1}$$

As the gas-liquid interface descends, the elevation difference between the highest point of the actual gas-liquid interface and the topmost perforation of the first row of oil

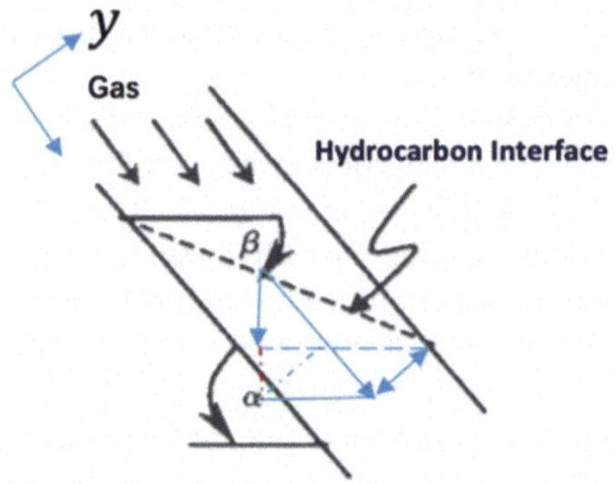

Fig. 4. Basic geometric relationship of Dietz model

wells and its position relative to the horizontal plane diminishes. This reduction in the included angle, as referenced in formula (1), leads to a decrease in the critical velocity required for maintaining interface stability.

That is to say, the reasonable gas injection velocity of gas injection gravity stable flooding changes with the interface movement. When the downward velocity (∂x/∂t) of the oil-gas interface is considered, the maximum included angle can be expressed as β:

$$\tan \beta_{max} = \mathrm{H} - \int_0^t \frac{\partial x}{\partial t} dt / D \quad (2)$$

From the insights gained through the physical simulation depicted in Fig. 1, it is observed that the leading edge of the gas-liquid interface becomes unstable when approaching the boundary of the pressure drop funnel associated with the production well. If the angle (β) is further reduced, the denominator in Eqs. (1) and (2) will be significantly influenced by the characteristics of the pressure drop funnel.

When examining the dynamics within a two-dimensional pressure drop funnel as outlined in reference [15] and accounting for the sedimentary rhythm of inclined reservoirs, we determine that permeability should be modeled as the harmonic average of vertical (K_v) and horizontal permeabilities (K_h). Given that the actual pressure drop funnel around oil wells manifests in three dimensions and acknowledging that the tangent function decreases as the angle diminishes, Eq. (2) must be formulated to adjust to these factors.

Thus, the equation could be restructured to incorporate these influences more accurately, possibly in a form such as:

$$\tan \beta_{max} = \frac{\mathrm{H} - \int_0^t \frac{\partial x}{\partial t} dt}{D + r_w e^{2\pi \sqrt{k_h k_v} \Delta p / (q_o \mu_o)}} \quad (3)$$

where: Δp is the production pressure difference, q_o is the oil production, and μ_o is the oil phase viscosity. If oil and water are produced simultaneously, $q_o \mu_o$ can be considered liquid production.

According to the Dietz geometric relationship in Fig. 4, the equivalent relationship of Dietz gas-liquid interface described by Gao Zhenhuan et al. [16], and the research of Mi Honggang and Lei Xiao, it is believed that the recovery factor of gas injection at the top of the reservoir with high dip angle of capillary force has little influence, and the recovery factor of gravity-assisted flooding for 15 years is only 0.2% [14], and the capillary force can be ignored, and if the interface is stable and the fluid velocities on both sides of the gas-liquid interface are equal, there are:

$$\tan\beta = \frac{u\left(\frac{\mu_o}{k_o} - \frac{\mu_g}{k_g}\right)cos\alpha}{g\Delta\rho_{o,g} - u\left(\frac{\mu_o}{k_o} - \frac{\mu_g}{k_g}\right)sin\alpha} \tag{4}$$

The equivalence relationship of formula (4) can also be written as:

$$\sin\alpha - \frac{dy}{dx}cos\alpha = \frac{u\left(\frac{\mu_o}{k_o} - \frac{\mu_g}{k_g}\right)}{g\Delta\rho_{o,g}} \tag{5}$$

where g is the acceleration of gravity, ko and kg are the permeability of the oil and gas phases, respectively, μ_o and are the viscosity of oil and gas, respectively, $\Delta\rho_{o,g}$ is the density difference between oil and gas.

According to the research of Xiong Yu et al. [9], for gas injection at the top of high-angle reservoir after water flooding, the oil phase permeability in formulas (4) and (5) can be taken as the oil phase permeability under the residual oil saturation of water flooding, while the gas phase permeability should be taken as the gas phase permeability under the residual liquid saturation after the gas-liquid interface.

According to the Dietz geometric relationship in Fig. 4, there are:

$$\tan\beta = \frac{dz}{dxcos\alpha + dysin\alpha} \tag{6}$$

where as according to Fig. 4, there is actually:

$$\mathrm{dz} = \mathrm{dxsin}\alpha - \mathrm{dycos}\alpha \tag{7}$$

Then formula (6) can be rewritten as formula (8):

$$\tan\beta = \frac{\mathrm{dxsin}\alpha - \mathrm{dycos}\alpha}{dxcos\alpha + dysin\alpha} \tag{8}$$

Equation (8) can be rewritten as Eq. (9):

$$\tan\beta = \frac{\mathrm{sin}\alpha - \frac{dy}{dx}\mathrm{cos}\alpha}{cos\alpha + \frac{dy}{dx}sin\alpha} \tag{9}$$

From Eqs. (5) and (9), it can be obtained:

$$\tan\beta = \frac{u\left(\frac{\mu_o}{k_o} - \frac{\mu_g}{k_g}\right)}{(cos\alpha + \frac{dy}{dx}sin\alpha)g\Delta\rho_{o,g}} \tag{10}$$

Combined with Eqs. (3) and (10), the maximum velocity form of reasonable gas injection gravity stable flooding considering the influence of injection-production well elevation difference, well spacing, reservoir rhythm and production well pressure drop funnel:

$$\mathrm{H} - \int_0^t \frac{\partial x}{\partial t} dt = \frac{u\left(\frac{\mu_o}{k_o} - \frac{\mu_g}{k_g}\right) D_e}{cos\alpha + \frac{dy}{dx} sin\alpha g \Delta\rho_{o,g}} \tag{11}$$

wherein:

$$D_e = D + r_w e^{2\pi\sqrt{k_h k_v}\Delta p/(q_o \mu_o)}$$

For an actual reservoir with secondary gas top gravity-assisted flooding, there must be multi-line effective wells. When the first-line effective well is shut-in or horizon adjustment is carried out after gas channeling, then the reservoir elevation difference H can be the elevation difference between the top gas injection well perforation section and the top perforation section of the edgemost oil well, that is, the maximum elevation difference or the closed height of the inclined reservoir.

In Eq. (11), when considering the stable movement of the gas-liquid interface, there must be dx = ∂x, and dy << dx, indicating that the unstable downward channeling velocity of the interface is significantly less than the overall downward moving velocity of the interface.

The downward moving velocity of the interface, as derived from material balance, can be mathematically expressed as

$$\frac{\partial x}{\partial t} = \frac{q_g B_g}{A} \tag{12}$$

where: q_g is the daily surface gas injection volume, B_g is the volume coefficient of the injected gas, and A is the average area of the gas cap.

Because the injection volume is relatively unstable, the cumulative injection volume can be considered to be expressed as the raw hydrocarbon volume injected into the ground (HCPV), denoted by the symbol B, as:

$$\mathrm{H} - \mathrm{B}/\mathrm{A} = \frac{u\left(\frac{\mu_o}{k_o} - \frac{\mu_g}{k_g}\right) D_e}{(cos\alpha + \frac{dy}{dx} sin\alpha) g \Delta\rho_{o,g}} \tag{13}$$

Therefore, the stable velocity expression considering the injection-production elevation difference, injection velocity, deposition rhythm and well spacing:

$$u_{st} = \frac{\mathrm{H} - \mathrm{B}/\mathrm{A}\,(cos\alpha + \frac{dy}{dx} sin\alpha)\, g \Delta\rho_{o,g}}{\left(\frac{\mu_o}{k_o} - \frac{\mu_g}{k_g}\right) D_e} \tag{14}$$

Because dy << dx when the gas-liquid interface moves downward steadily. Therefore, the ratio of stable velocity (u_{st}) to critical velocity (u_c) of the artificial gas cap in the inclined reservoir can be approximated as:

$$\frac{u_{st}}{u_c} = \frac{H - B/A}{D_e tan\alpha} \tag{15}$$

Obviously, the increase of oil recovery rate will lead to the increase of production pressure difference, the worse stability of gas-liquid interface and the smaller the required stable speed. When the oil field gas injection rate is constant and the downward movement of gas-liquid interface, the stable speed will gradually increase. Formula (15) conforms to the basic physical process of gravity stable oil flooding of artificial gas cap in actual inclined reservoirs. Obviously, gas injection gravity flooding stabilization velocity is not a widely used J. M. The method shown in Eq. (16) given by Dumore method [2]. The method shown in formula (16) is also adopted in the current national standard of gas injection gravity stable flooding in China. Of course, it is absolutely stable, but it is not in line with the output pursued at the site and the engineering requirements.

$$u_{st} = \left(\frac{d\rho}{d\mu}\right)_{min} kg \tag{16}$$

where k is the absolute permeability of the reservoir.

When the structure of the high-angle reservoir is incomplete, that is, the boundary of the reservoir is a part of the complete structure, if the angle radian between the top and the boundary on both sides of the reservoir is assumed to be θ, the average effective reservoir thickness is h, as shown in Fig. 5. From the geometric relationships such as the oblique thickness of the stable gas-liquid interface and the length difference between the inside and the outside of the same contour line, it can be seen that the gas displacement volume velocity when the interface moves downward has the following relationships:

$$\frac{dV}{dt} = \pi\theta(h/sin\alpha)^2 \frac{\partial x}{\partial t} \tag{17}$$

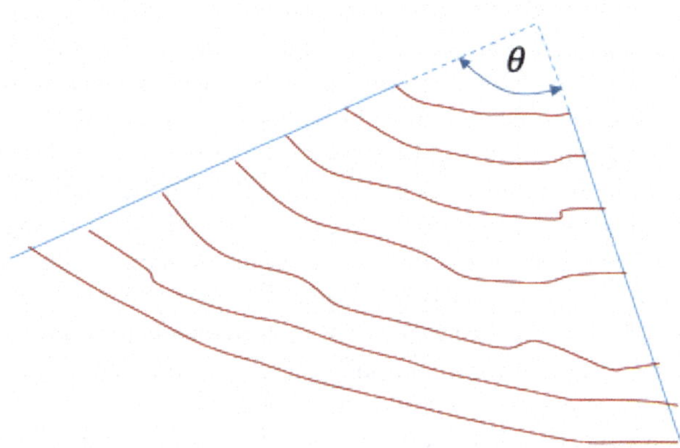

Fig. 5. Schematic diagram of oil reservoir clamped by high dip fault

Therefore, it is easy to know that the basic relationship between the stable speed and the critical speed can be expressed as:

$$\frac{u_{st}}{u_c} = \frac{H - B/(\pi\theta(h/sin\,\alpha)^2)}{D_e tan\,\alpha} \tag{18}$$

3 Case Analysis and Application

3.1 Basic Situation of Example Reservoir

The LB gas injection reservoir is a fault descending disk nose structure, the formation dip angle is 30° ~ 50°, the oil-bearing area is 1.4 km^2, the buried depth of the reservoir is-2700 ~ −3200 m, the reservoir temperature is 93 ~ 113 °C, and the average temperature is 101.8 °C.

The reservoir is dominated by fan delta and braided river delta braided channel, with an average porosity of 19.5%, an average horizontal permeability of 265 × 10–3 μm2, an average vertical permeability of 191 × 10–3 μm2, and a vertical/horizontal permeability of 0.77. The oil product is conventional light oil, the density of underground crude oil is 0.78 g/cm3, the density of surface crude oil is 0.85 g/cm3, and the viscosity of underground crude oil is 1.73. The top gas injection test began on October 24, 2022.

3.2 Dynamic Analysis of Actual Gas Injection

The test injection wells are LBJ1, LBJ2, LBJ3 and LBJ4 from NE to SE, all of which are vertical wells. During test injection, due to physical properties, phase change and being too close to fault, LBJ1 had less injection volume. Well LBJ2 had opposite oil production wells along the downdip direction of the structure, but wells LBJ3 and LBJ4 had no oil production wells. Moreover, the elevation difference and well spacing of oil production wells in the downdip direction of LBJ2 were the smallest. Due to the influence of surface engineering, injection will be stopped in February 2024. Figure 6 shows the water cut and oil production dynamics after gas injection in the block. The gas injection effect is obvious, but during the test injection process, well LBJ2 is facing the downward direction in March 2023. Gas channeling occurs in the oil well. At this time, there is no gas channeling in wells LBJ3 and LBJ4 because the spacing between production wells in downdip direction is larger than that of LBJ2. Obviously, the stability of gas-liquid interface is affected by the above-mentioned factors such as elevation difference and well spacing between injection and production wells.

As can be seen from Fig. 9, the first gas injection well group with gas channeling is LBJ2 well group, which is obviously consistent with the aforementioned injection-production ratio affecting the stability of gas-liquid interface of gravity stable flooding. At the same time, the field well test interpretation confirms that there is a high permeability rhythmic layer between the layers of LBJ2 well group (Fig. 7).

Using the ratio relationship between stable velocity and critical velocity of artificial gas cap in inclined reservoir established in this paper, the critical velocity, stable velocity, gas-liquid interface downward velocity and sweep area of LBJ2 well group are

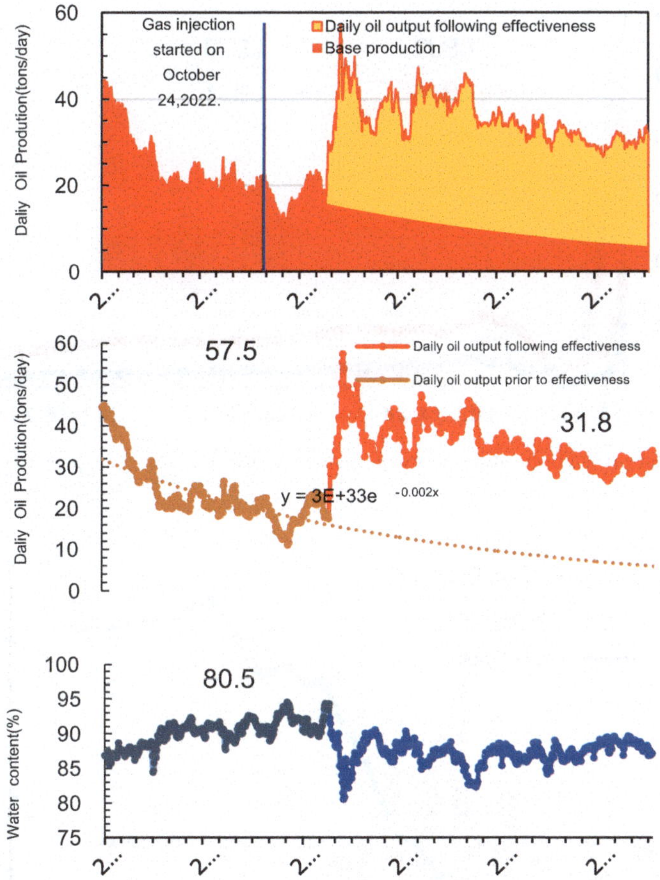

Fig. 6. Dynamic curve of gas injection in high dip angle reservoir of LB reservoir

calculated, as shown in Fig. 8. The critical velocity algorithm is calculated by using the prediction method given by Xiong Yu et al. [9] considering the residual oil condition after water flooding and the residual liquid condition after gas flooding.

According to the method in this paper, the critical gas injection velocity of LBJ2 well group is 2.2328 m/a, and the stable velocity of gravity stable gas-liquid interface is 1.0835 m/a. It can be seen from Fig. 10 that the interface migration velocity of this well group reaches the maximum from the end of February to the beginning of this month 2023, which is consistent with the gas channeling time of this well group; In the later period, when no measures were taken for gas channeling wells, the gas-liquid interface migration velocity was greater than the stable velocity most of the time. Then, after the gas channeling wells were suppressed, the gas-liquid interface migration velocity gradually fell between the stable velocity and the critical velocity. From November 2023 to December 2023, the gas-liquid interface migration velocity of this well group was lower than the stable velocity of gravity-assisted flooding after the injection was stopped for 2 months and after the gas channeling wells were treated. After the injection is

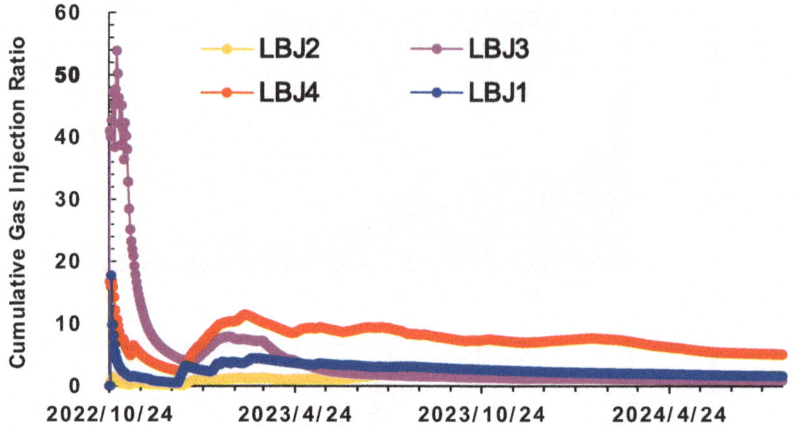

Fig. 7. Injection-production ratio curves of each test injection well group in LB gas injection reservoir

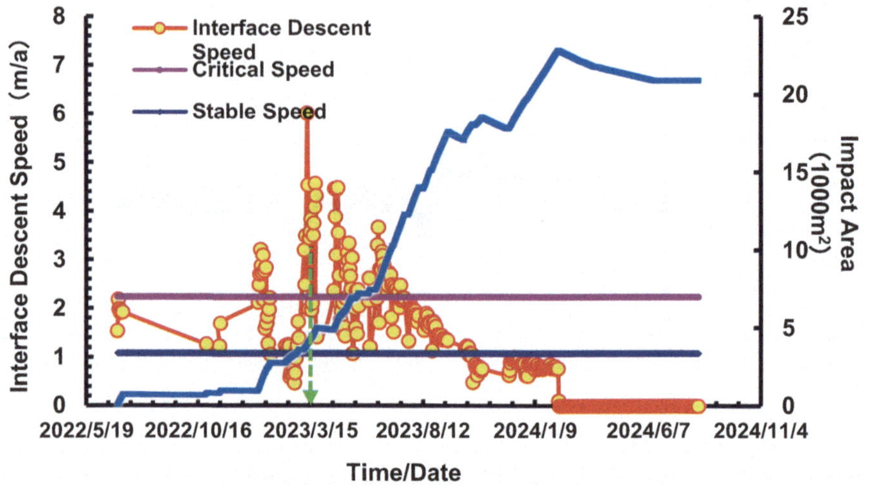

Fig. 8. Comparison of LBJ2 gas injection velocity with critical velocity and stable velocity

completely stopped in 2024, the affected area of this well group will shrink due to the inability of gas channeling wells to be completely shut in.

The above-mentioned critical velocity and stable velocity correspond to the annual injection velocity of 5.45% HCPV and 2.64% HCPV respectively, which are lower than the designed injection velocity of 0.06 HCPV/a, indicating that in order to achieve complete gravity stable flooding in LB reservoir, the top area is small in the early stage of gas cap formation, and the designed injection velocity is high.

4 Conclusion

(1) The prediction method of stable velocity of gravity-assisted flooding of gas injection established in this paper can correctly reflect the migration law of gas-liquid interface of top gas injection;
(2) Considering the production requirements, the stable velocity of gas injection gravity-assisted flooding is larger than the absolute stable velocity, which is a function of engineering factors and geological factors such as elevation difference between injection and production wells, well spacing and injection-production ratio.
(3) The design injection velocity of gravity stable flooding of gas injection in LB reservoir is higher than the stable velocity, so the gas channeling well should be closed to promote the formation of gas cap and realize gravity-assisted stable flooding.

References

1. Dietz, D.N.: A theoretical approach to the problem of encroaching and bypassing, edge water. Konikl. Ned. Akad. Wetenschap, Proc. B56, 83, 1953
2. Dumore, J.M.: Stability considerations in downward miscible displacements. SPE J. **9**, 356–362. (1964). SPE 961
3. Perrine, R.L.: Stability theory and it uses to optimize solvent recovery of oil. SPE J. 9–16 (1961). SPE 1508-G
4. Sheldon, J.W., Fayers, F.J.: The motion of an interface between two fluids in a slightly dipping porous medium. SPE J. 275–282 (1962). SPE 325
5. Dougherty, E.L.: Mathematical model of an unstable miscible displacement. 155–163 (1963). SPE 509
6. Cardenas, R.L., Alston, R.B., Nute, A.J., et al.: Laboratory design of a gravity-stable miscible CO_2 process. J. Pet. Technol. 111–118 (1984). SPE 10270
7. Ypma, J.G.J.: Analytical and numerical modeling of immiscible gravity-stable gas injection into stratified reservoirs. SPE J. 554–564 (1985). SPE12158
8. Fayers, F.J., Muggerldge, A.H.: Extensions to dietz theory and behavior of gravity tongues in slightly tilted reservoirs. SPE Reservoir Engineering, November 1990: 487–494. SPE 18438
9. Xiong, Y., Liangtian, S., Lei, S., et al.: Study on reasonable gas injection velocity of N2 immiscible flooding in inclined multi-layer reservoirs. J. Southwest Pet. Inst. **24**(5), 34–36 (2002)
10. N.A.F. El-Khatib. The Modification of the Dykstra-Parsons Method for Inclined Stratified Reservoirs. Sep. 2011, SPE 140960
11. Shaoran, R., Yanmin, L., Liang, Z., et al.: Discriminant model and test of gravity stabilized gas flooding. J. China Univ. Pet. (Nat. Sci. Edit.) **42**(4), 59–66 (2018)
12. Rui, W.: Stable gas flooding conditions and influencing factors of artificial gas cap. Sci. Technol. Eng. **21**(24), 10236–10245 (2021)
13. Hameed, M., Konwar, L., Alanaisi, I., et al.: Reservoir Performance Analysis of Eighty-Four Years of Immiscible Gas Injection in a Matured Carbonate Reservoir, Bahrain Field. October 2022, SPE-211348-MS
14. Honggang, M., Xiao, L.: Research on gravity-assisted stable displacement mechanism of gas injection in Weizhou 12–1 oilfield. Pet. Drill. Prod. Technol. **29**(6), 28–31 (2007)
15. Kong, X.: Advanced Seepage Mechanics. China University of Science Press, 1999
16. Gao, Z., Liu, Z., Du, X.: Oilfield gas injection and production technology. Petroleum Industry Press, 1994

Impact of Solar Greenhouse Attachment on Indoor Thermal Comfort and Air Quality in Rural Dwellings During Winter in Severe Cold China

Xueyan Zhang(✉), Bingyang Wei, and Kaibiao Wang

Laboratory of Building Environment and New Energy Resources, Dalian University of Technology, Dalian 116024, Liaoning, China
xueyan@dlut.edu.cn

Abstract. Frequent occurrences of smoggy weather are increasingly raising people's requirements for indoor environments. Not only is thermal comfort indoors important, but the diffusion characteristics of indoor particulate matter must also be considered. In rural residences in severely cold regions, when farmers engage in activities such as cooking and heating, high concentrations of particles released from stoves and furnaces can spread throughout the room and adjacent rooms. Additionally, the temperature differences created by heating methods such as brick beds (kang), soil-based heaters, and stoves can lead to natural convection within the house. Particles are influenced by the indoor temperature gradient and move within the indoor space. Therefore, it is practically significant to investigate how rural residents' activities and insulation methods affect the indoor thermal environment and the diffusion characteristics of particles. This study conducted experiments in a rural residence in Shenyang City. The results show that applying plastic film to the exterior walls as an insulation method can effectively improve the temperature distribution in the horizontal direction indoors and significantly reduce the temperature non-uniformity coefficient, while having no noticeable effect on temperature distribution in the vertical direction. Cooking and heating activities in rural residences significantly disturb the concentration and distribution of PM2.5 and PM10 indoors. Fine particles remain in the central area of the room and in the breathing zone of occupants for extended periods due to thermophoresis. Among different fuels, coal and straw combustion generate the most pollutants, which tend to linger in the kitchen for long periods.

Keywords: Rural dwellings · Severe cold China · Non-uniformity heat transfer · Thermophoresis

1 Introduction

As living standards among farmers improve, so do the requirements for their living environments. Creating healthy and comfortable living spaces and enhancing the quality of life are becoming increasingly important concerns for rural residents. In severely cold

C. Yuan (Ed.): ICRER 2024, ESE, pp. 132–143, 2025.
https://doi.org/10.1007/978-3-031-93295-3_11

regions, the significant temperature difference between indoors and outdoors leads rural households to adopt corresponding wall insulation measures to cope with the harsh outdoor conditions. Due to the influence of residents' living habits and economic constraints, some opt for transparent insulation measures, such as attaching plastic films to exterior walls, to enhance the airtightness of building envelopes. Such measures using plastic film are simple to implement and cost-effective; however, while improving airtightness for better insulation, they may also impact the rate at which indoor particulate matter is expelled. During cooking and heating periods, stoves and heating furnaces emit large amounts of inhalable particulate matter into the kitchen, which rapidly spreads throughout the house and to adjacent rooms, posing health risks to the inhabitants. Studies have shown that exposure to PM2.5 is associated with increased incidences of premature birth and asthma [1, 2]. Furthermore, the World Health Organization reports that, based on toxicological and clinical medical studies, short-term exposure to particulate matter produced by solid fuel combustion can immediately cause physiological changes in the human body [3]. Huang Guanhua et al. [4] conducted field tests and tracking records to clarify the activity patterns of residents inside rural homes, assessing the PM2.5 exposure levels and health risks faced by indoor occupants, finding that the health risks and economic losses caused by pollution cannot be ignored.

Research indicates that heating activities in rural homes are the primary reason for the higher concentration of PM2.5 in various rooms [5]. Limited by economic conditions, rural homes use heating facilities like brick beds (kang), soil-based heaters, and stoves, leading to uneven heat distribution and significant temperature differences within the home. Under the influence of indoor temperature gradients, particles move through the indoor space. Li [6] found through theoretical research and numerical simulations that thermophoresis is one of the main forces affecting the movement of submicron aerosol particles. The smaller the particle size, the higher the heat source temperature, and the lower the initial velocity of the particles, the greater the dispersion of indoor particles, and the more intense the movement of aerosol particles becomes. Xiujuan Wang [7] discovered through the study of the effects of thermophoresis on the movement characteristics of indoor particles that in areas with large temperature gradients near cold walls, particles exhibit a high concentration distribution phenomenon. However, the deposition rate on horizontal hot walls is significantly reduced; when the wall temperature exceeds the fluid temperature by about 20°C, smaller particles find it difficult to deposit on the wall surface [8]. Numerous scholars, including McLaughlin have demonstrated through CFD simulations and experimental methods that the phenomenon of thermophoresis plays a significant role in the deposition of particles on surfaces and their distribution within indoor spaces. Additionally, the heating method using radiators significantly affects the distribution of indoor temperatures and the dispersion of solid particles [9–11].

According to the survey results of winter indoor temperature measurements in rural residences in Liaoning, the indoor temperature of these residences fluctuates dramatically due to low outdoor temperatures and the heating from stoves and kang (brick bed). Overall, the temperature in bedrooms and living rooms varies between 1 °C and 25 °C, while the temperature in kitchens and utility rooms is lower than in bedrooms, generally ranging from 5 °C to 15 °C [11–14]. During heating and cooking activities, there are significant temperature and concentration differences at different locations within the rooms. It is evident that rural residences in cold regions are strongly influenced by outdoor environmental conditions and heating methods such as stoves and soil-based heating systems, resulting in significant temperature gradients both within the same room and between different rooms. The thermophoretic phenomena in these environments play a non-negligible role in the diffusion and deposition of particles. Therefore, it is necessary to conduct experimental studies on the movement characteristics of particles under non-uniform temperature gradients within rural residences.

2 Experiments and Methods

2.1 Experimental Conditions

The experimental site was located within a rural residence in Datai Village, Shenyang City, Liaoning Province, as shown in Fig. 1. The residence is symmetrical from east to west and mainly consists of a kitchen and a bedroom. The primary experimental instruments were placed in the kitchen and bedroom on the eastern side of the residence, and the experiment involving laying plastic film on the outer windows was conducted in the kitchen and bedroom on the eastern side. The layout of the measurement points for the indoor instruments are illustrated in Fig. 2. All instruments were set to continuous monitoring mode, recording data every 1min. The distance of the temperature and humidity measurement points from the south wall in the bedroom was 0.6 m, 2.4 m, and 3.0 m, respectively. In the kitchen, the distances of the temperature and humidity measurement points from the south wall were 0 m, 1 m, and 2 m, respectively, with each temperature and humidity measurement point having a corresponding PM2.5measurement point. The height of the measurement points in the central area of the kitchen was set at 0 m and 1.2 m; near the inner wall, the heights of the measurement points were set at 0 m, 1.2 m, and 2.4 m. In the bedroom, the heights of the temperature and humidity measurement points were all set at 0.6 m, 1.2 m, and 1.5 m. The PM2.5measurement points were located at position B, with heights of 0.6 m, 1.2 m, and 1.5 m.

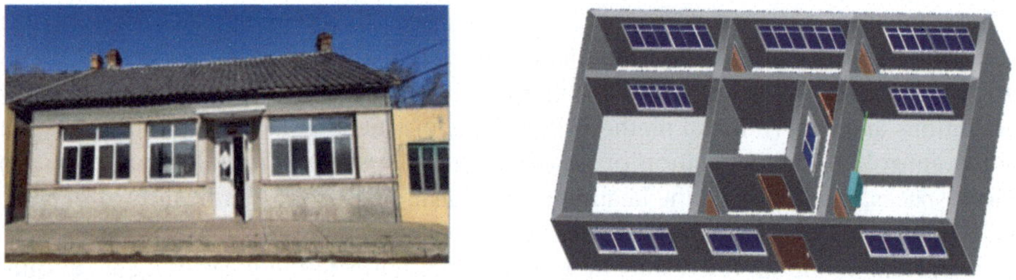

(a) Building façade (b) Layout diagram

Fig. 1. Interior structure of rural house

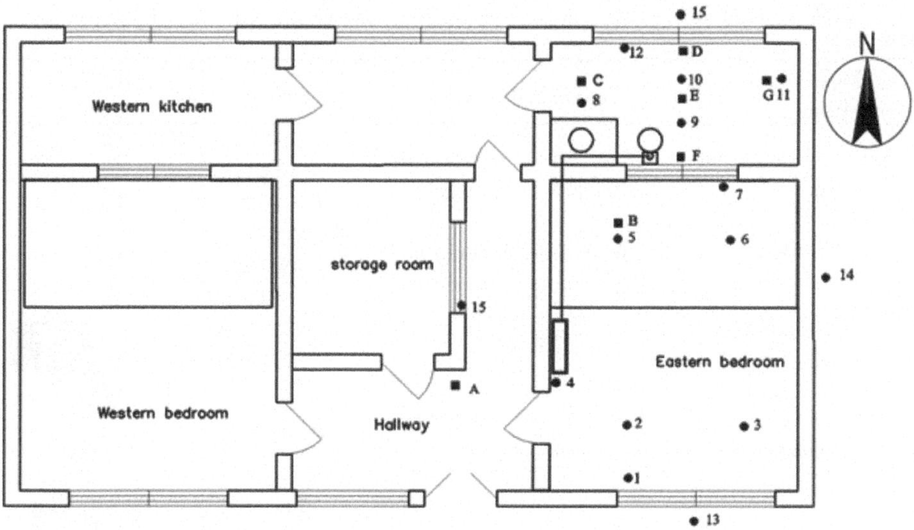

Fig. 2. Indoor measurement point layout diagram

2.2 Experimental Methods

To closely simulate the actual conditions of the rural residence, during the experiment, coal was added to the stove starting at 9:00 AM to provide heating for the kitchen and bedroom. At 9:00 AM, 12:00 PM, 3:00 PM, and 6:00 PM, different fuels such as wood, coal, and straw were added to the stove to investigate the impact of cooking activities with different fuels on indoor temperature and humidity as well as the distribution of indoor particulate matter.

The instruments for measuring temperature, humidity, and wall heat flux consist of 15 temperature and humidity loggers, 1 WZY-2 temperature logger, 1 RLZY-1 heat flux recorder, and 2 multi-channel heat flux recorders. Among these, the two multi-channel heat flux recorders measure the heat flux and surface temperature of the walls in the bedroom and kitchen. The temperature and humidity loggers and particulate matter measurement instruments monitor the spatial distribution of temperature, humidity, PM2.5, and PM10 in the kitchen, bedroom, and hallway.

Additionally, to study the impact of applying plastic film to the exterior of the rural residence on indoor temperature, humidity, and particulate matter distribution, on the second day of the experiment, the south exterior window of the bedroom and the north exterior window of the kitchen were sealed with plastic film. The experiments continued in the kitchen and bedroom, recording changes in the indoor environment. The environmental measurement instruments used inside the rural residence are listed in Table 1.

In this experiment, the main fuels used were charcoal, corn cobs, coal, and straw. The outer windows of the bedroom and kitchen were sealed with plastic film to investigate the impact of plastic film on indoor temperature and humidity, as well as the effect of different fuels on the distribution of indoor particulate matter. According to the survey results

Table 1. Farm house environmental parameters measuring instrument parameters

Measurement Parameters	Testing Instruments	Instrument Accuracy	Instrument Accuracy
Air Temperature and Humidity	WSZY-2-Temperature and Humidity Data Logger	-40~100°C (±0.1°C) 0~100%RH (±0.1%RH)	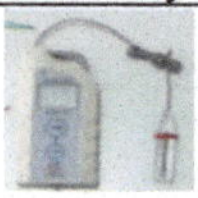
Air Flow Velocity	WFWZY-1-Air Velocity and Temperature Data Logger	-20~80°C (±0.5°C) 0.05~30m/s (5%±0.05m/s)	
PM10 and PM2.5 Concentrations	ZF-R3	0~3000μg/m³ (±1μg/m³)	
CO_2 Concentration	WEZY-1-Carbon Dioxide Data Logger	0~5000ppm	
Wall Temperature and Heat Flux	Multi-Channel Heat Flux Meter	-50~120°C (±0.1°C) 0~2000 W/m² (±0.1W/m²)	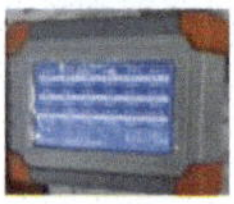
Wall Heat Flux Density	RLZY-1	-50~100°C (≤5%)	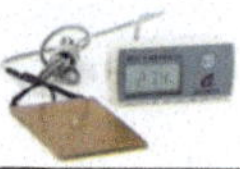

of winter indoor temperature measurements in rural residences in Liaoning, the indoor temperature of these residences fluctuates dramatically due to low outdoor temperatures and the heating from stoves and kang (brick bed). Overall, the temperature in bedrooms and living rooms varies between 10 °C and 25 °C, while the temperature in kitchens and utility rooms is lower than in bedrooms, generally ranging from 5 °C to 15 °C [11]. Moreover, during heating and cooking activities, there are significant temperature and concentration differences at different locations within the rooms. Therefore, experimental methods can be used to preliminarily explore the movement of indoor particulate matter.

The experimental conditions and procedures for this experiment are as follows:

(1) On the day before the experiment, the instruments in the eastern kitchen and eastern bedroom of the rural residence were cleaned and calibrated. Temperature and humidity loggers, anemometers, ZF-R3 type loggers, and multi-channel heat flux meters were set up in the experimental rooms to monitor the environment without any heating measures.
(2) On the first day of the experiment, the cooking conditions in the kitchen and bedroom were established, and the stove was lit at 8:00 AM to provide heating for the radiators in the kitchen and bedroom. The fuel for the heating stove was coal, and the heating lasted for the entire day. The cooking activities used wood and corn cobs as fuel. Changes in the indoor environment of the bedroom, kitchen, and hallway were recorded throughout the day.

(3) On the second day, while maintaining the heating and cooking conditions, plastic film was applied to the south exterior window of the bedroom and the north exterior window of the kitchen to enhance the airtightness of the building envelope. The fuel for the heating stove remained coal, and the heating lasted for the entire day. The cooking activities used coal and corn cobs as fuel. Additionally, the solar radiation intensity at the exterior windows of the eastern and western bedrooms was recorded at several time points in the morning

(4) On the third day, normal cooking and heating activities continued under the condition of plastic-sealed windows. The cooking activities used straw and corn cobs as fuel. After the experiment concluded on the fourth day, the instruments were collected and organized, and the experimental data were exported.

3 Experimental Results and Analysis

3.1 Temperature Distribution Changes Inside the Rural Residence

Figure 3 and Fig. 4 show the outdoor temperature changes of the rural residence from January 15 to January 18, along with the temperature distribution inside the bedroom at 4:00 AM. During this period, there was no heating in the bedroom, and no one was moving around inside, allowing the temperature and humidity distribution to truly reflect the impact of the plastic film sealing windows on the thermal and moisture environment inside the rural residence under natural conditions. In the experiment, the indoor environment on January 15 and 16 was without the plastic film, while the data on January 17 and 18 was after the plastic film was applied. From the figures, it can be observed that on January 15 and 16, without the plastic film, the temperature near the exterior window was 4 °C lower than the indoor temperature, and within a horizontal distance of 1.5 m from the exterior window, the indoor temperature rapidly dropped below 0 °C. On January 18, the temperature distribution inside the bedroom with the plastic film applied is shown, and at this moment, the outdoor temperature differed by less than 2 °C compared to January 15. Although the temperature near the window was still the lowest indoors, the temperature drop near the window was not as pronounced, and the temperature distribution was more uniform. On January 17, due to higher outdoor temperatures, the indoor temperature was noticeably higher than on other experimental days. It is evident that, in the absence of solar radiation, the transparent insulation measures of the exterior windows of the rural residence can, to some extent, improve the uneven temperature distribution inside the residence and enhance indoor comfort.

The temperature distribution inside the room after applying plastic film to the exterior walls can be evaluated using the "temperature non-uniformity coefficient". To obtain the temperature non-uniformity coefficient inside the room, five points are selected in each room, and the temperature non-uniformity coefficient under different conditions is calculated using Eq. (1).

$$\sigma_T = \sqrt{\frac{1}{n}\sum_{i=1}^{n}\left(\frac{(T_i - T_0)^2}{T_0}\right)} \tag{1}$$

In the equation, σ_T is the temperature non-uniformity coefficient; n is the number of measurement points; T_0 is the average temperature (K); T is the temperature value at each

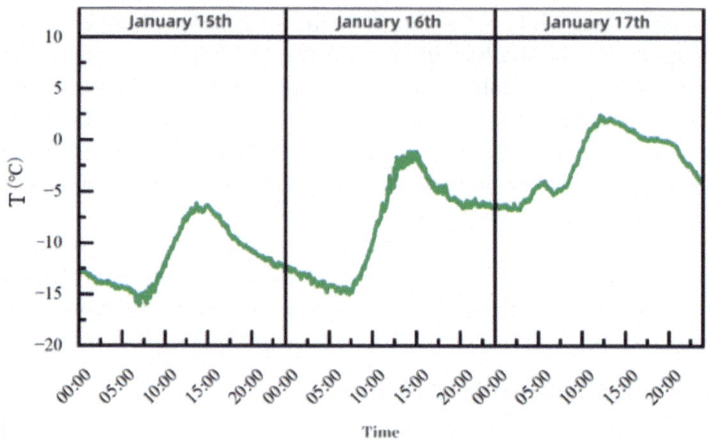

Fig. 3. Bedroom interior cloud map axis and outdoor temperature change

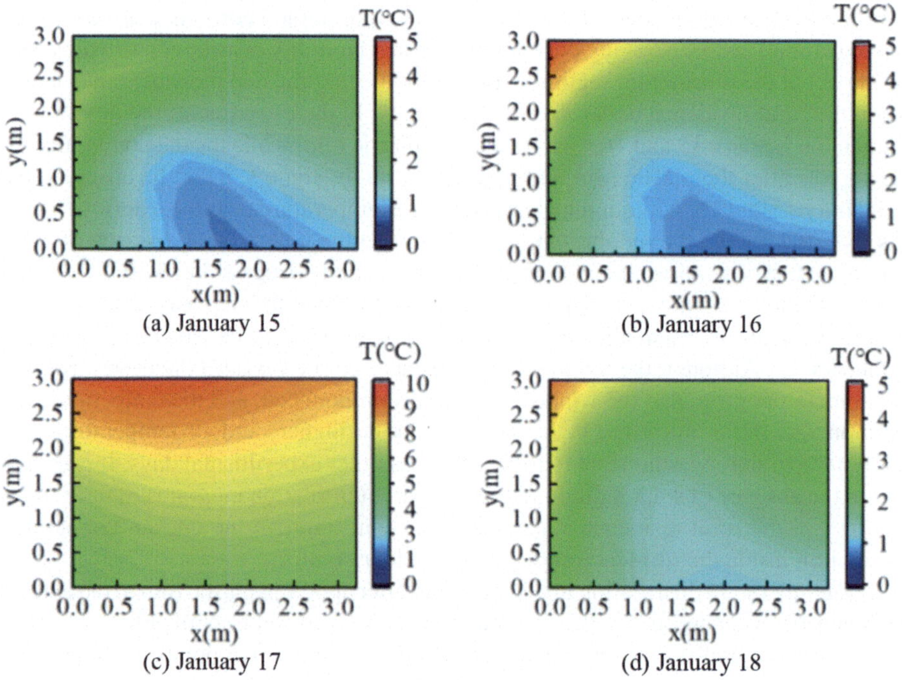

Fig. 4. Temperature field distribution in bedroom

measurement point (K). The smaller the σ_T, the more uniform the indoor temperature distribution.

After calculation, it is known that the temperature non-uniformity coefficients inside the bedroom at 4:00 AM on January 15, 16, 17, and 18 were 63%, 60%, 23%, and 34%, respectively. This indicates that applying plastic film to the exterior windows reduced the indoor temperature non-uniformity coefficient by approximately 30%.

Figure 5 shows the temperature variation trends at different positions, measured at heights of 1.5 m and 1.0 m, from the south exterior wall of the bedroom from January 15 to January 18. From the figure, it can be observed that regardless of whether plastic film is applied to the exterior windows, the temperature differences at different heights within the space are minimal, with typical temperature fluctuations being less than 1 °C. Only during midday and afternoon when solar radiation is stronger, the temperature at different heights suddenly increases.

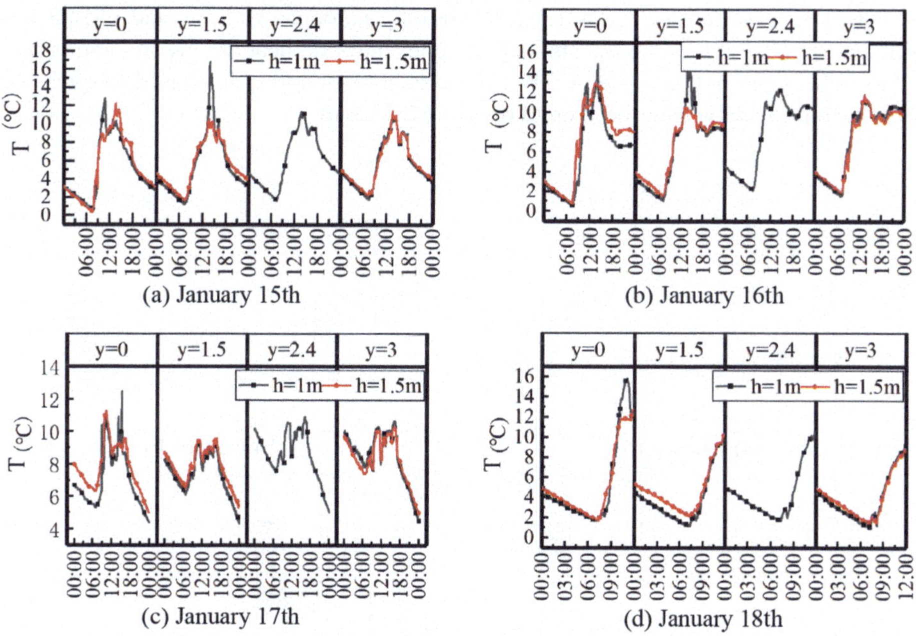

Fig. 5. Bedroom horizontal and vertical temperature distribution

The temperature at points near the north interior window remains almost constant along the height. This is because the deeper areas inside the room are less affected by the south exterior window and are closer to the heat source. This indicates that the application of plastic film to the exterior windows has a consistent effect on the temperature at different heights in the areas where people are active within the bedroom. Additionally, it can be seen that the use of radiators and kang (brick bed) for heating in the rural residence primarily affects the horizontal temperature distribution within the residence, especially when the external air temperature is very low. However, at the height where people are active, the air temperature is very similar. This is due to the spatial layout of rural residences in Northeast China, where radiators and kang are typically placed near the interior walls.

3.2 Particle Diffusion Characteristics

Figure 6 shows the distribution of PM2.5 near the kitchen heat source and provides the PM2.5 mass concentrations at three different heights over a period after the fuel is ignited. By comparing the vertical distribution of PM2.5 at different times, it is observed that after the cooking activities begin, the highest PM2.5 concentration on January 17 was generated by burning straw. One and a half hours after the experiment started, the indoor concentration remained between 2000–3200 μg/m^3. The PM2.5 concentrations at different heights varied significantly, with PM2.5mainly concentrated at h = 2.4 m and near the ground, while the concentration in the breathing zone of human activity was still around 2000 μg/m^3.

When burning corn cobs, the relative humidity indoors was 57%, and PM2.5 was still concentrated at h = 2.4 m. The particle concentration near the ground and at h = 1.2 m differed by only 152 μg/m^3. When the relative humidity increased to 69%, the particle concentration near the ground rose to 1752 μg/m^3.

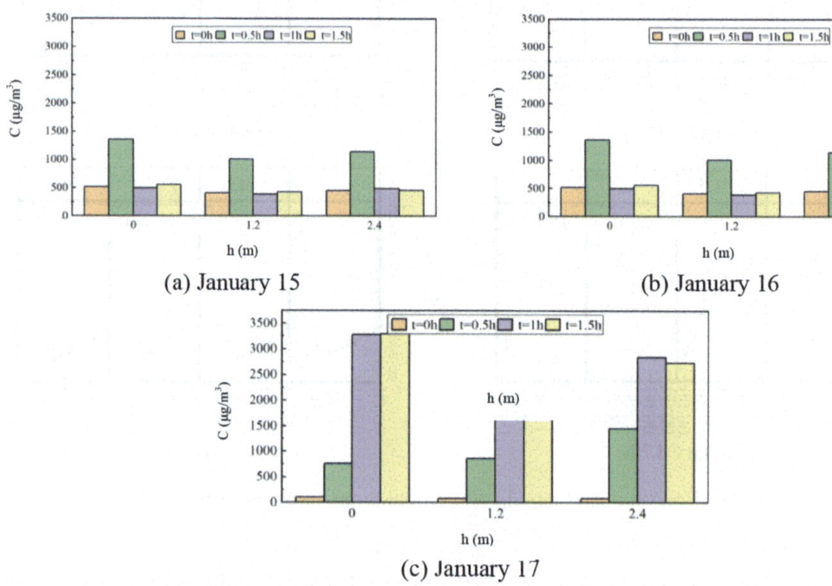

Fig. 6. Vertical distribution of PM2.5 inside farm house at different times

Particles of different sizes exhibit distinct movement characteristics in the air. Smaller particles are less affected by gravity, allowing them to remain suspended in the air for longer periods and diffuse more easily. Larger particles, due to gravitational forces, settle to the ground more quickly.

Figure 7 shows the concentration changes of PM2.5 and PM10 particles near the stove in the kitchen when burning four different fuels: coal, straw, corn cobs, and wood. From the figure, it can be observed that when no fuel is burned, the concentrations of PM10 and PM2.5 particles indoors are very close, with a difference of less than 100 μg/m^3. During the early morning hours when there is no human activity in the rural residence, the concentrations of PM2.5 and PM10 are between 70–105 μg/m^3, with a concentration difference of 10–20 μg/m^3. This indicates that the movement and handling of fuel by indoor occupants also affect the fluctuation of indoor particle concentrations, ranging from 130–600 μg/m^3 before the fuel is ignited. During the fuel combustion process, coal and straw produce higher particle concentrations that persist for longer durations. Within 2 h of the experimental conditions, the particle concentrations from wood and corn cobs return to the pre-experiment levels, while the indoor pollutant concentrations from coal and straw remain between 1200–2200 μg/m^3.

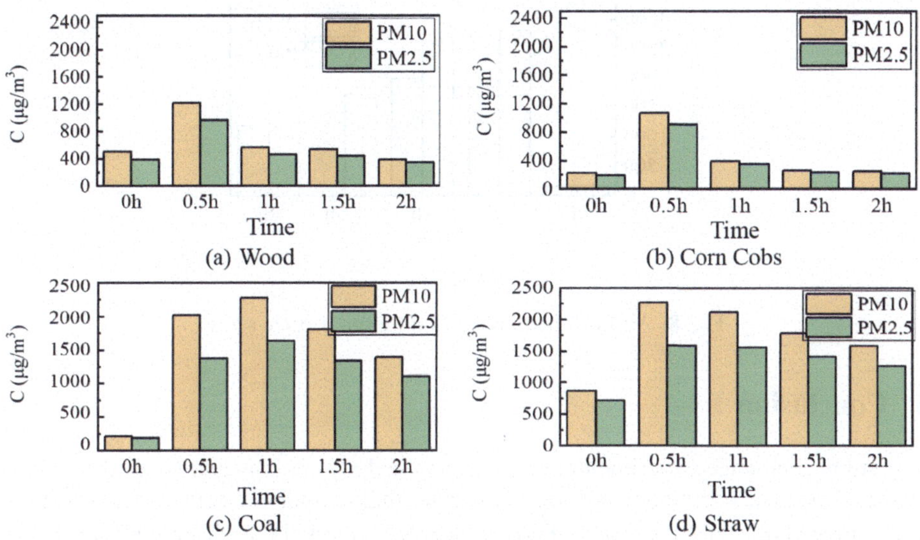

(a) Wood (b) Corn Cobs (c) Coal (d) Straw

Fig. 7. PM10 and PM2.5 concentrations from combustion of different fuels

Figure 8 shows the changes in particle concentration at different heights within the room over a certain period after fuel combustion. From the figure, it can be observed that near the heating equipment, the movement of particles of both sizes is primarily influenced by thermal buoyancy. Under the influence of thermal buoyancy, particles concentrate at a height of h = 2.4 m, with the highest concentration reaching 2752 μg/m^3 at this height, while the concentration is lowest at the intermediate measurement point at h = 1.2 m. Near the heating equipment, the movement of particles of both sizes is mainly influenced by thermal buoyancy.

The PM10 concentration generated by burning coal is higher. The concentration of larger particles decays more rapidly at h = 2.4 m, decreasing to 1238 μg/m^3/h one hour after the start of the experiment. At heights of 0.1 m and 1.2 m, the PM10 concentration decays to 538 μg/m^3/h and 507 μg/m^3/h, respectively. In contrast, the concentration of smaller particles (PM2.5) decays more slowly at heights of 0.1 m, 1.2 m, and 2.4 m, with decay rates of 357 μg/m^3/h, 295 μg/m^3/h, and 801 μg/m^3/h, respectively. These smaller particles remain in the human activity area for a longer period due to their slower decay rate.

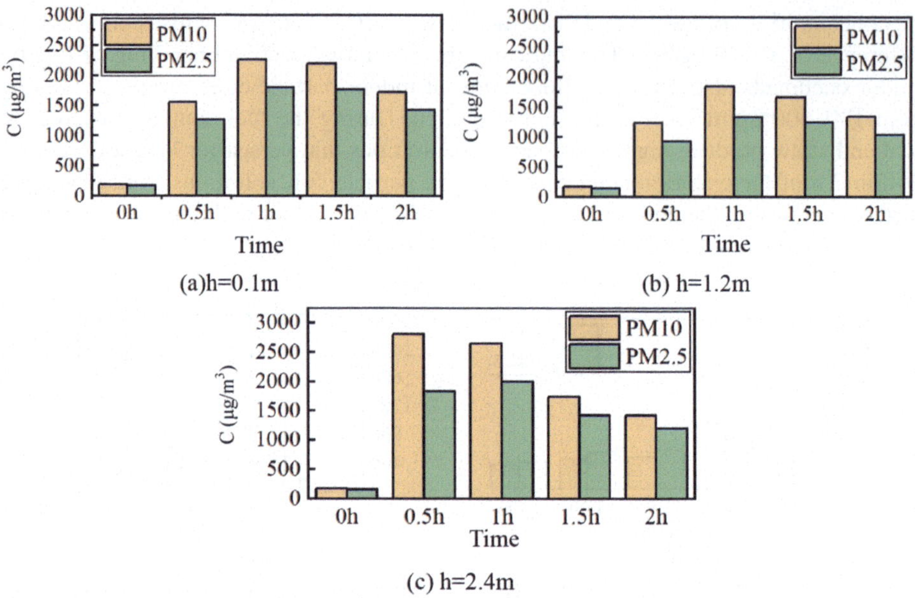

Fig. 8. Vertical distribution of different particle sizes

4 Conclusion

This study conducted experimental research on the daily cooking and heating activities in a rural residence in Shenyang. By comparing the thermal environment and particle distribution characteristics inside the residence with and without plastic film insulation on the exterior windows, the following conclusions were drawn.

(1) The application of plastic film on the exterior windows of the rural residence can, to some extent, improve the uneven temperature distribution inside the residence and enhance indoor comfort. Without solar radiation, applying plastic film to the exterior windows can reduce the horizontal temperature gradient indoors, while having almost no effect on the vertical temperature gradient. During the midday period when sunlight directly shines through the windows, the temperature inside the plastic film rapidly increases, creating a small greenhouse effect between the film and the window.

(2) The temperature gradient near the heating equipment is significant. PM2.5 and PM10 particles near the heating equipment are primarily influenced by thermal buoyancy and thermophoresis. Under the influence of thermal buoyancy, particles concentrate at a height of h = 2.4 m, which is close to the breathing zone of human heads. The PM10 concentration generated by burning coal is higher, and the concentration of larger particles decays more rapidly at h = 2.4 m, decreasing to 1238 $\mu g/m^3/h$ one hour after the start of the experiment. At heights of 0.1 m and 1.2 m, the PM10 concentration decays to 538 $\mu g/m^3/h$ and 507 $\mu g/m^3/h$, respectively. In contrast the concentration of smaller particles (PM2.5) decays more slowly at heights of 0.1 m, 1.2 m, and 2.4 m, with decay rates of 357 $\mu g/m^3/h$, 295 $\mu g/m^3/h$, and 801 $\mu g/m^3/h$,

respectively. These smaller particles remain in the human activity area for a longer period due to their slower decay rate.

Acknowledgments. This work was supported by the National Nature Science Foundation of China (No. 52078098 and No. 51608092), and National Science Foundation of Liaoning Province (No. 2019-ZD-0022).

References

1. Malley, C.S., Kuylenstierna, J.C., Vallack, H.W., et al.: Preterm birth associated with maternal fine particulate matter exposure: a global regional and national assessment. Environ. Int. **101**(1), 173–182 (2017)
2. Anenberg, S.C., Henze, D.K., Tinney, V., et al.: Estimates of the global burden of ambient PM2.5, Ozone, and NO2 on asthma incidence and emergency room visits. Environ. Health Perspect. **126**(10), 1–14 (2018)
3. Malley, C.S., Ashmore, M.R., Kuy lenstierna, J., et al.: Microenvironmental modelling of personal fine particulate matter exposure in Accra, Ghana. Atmos. Environ. **225**(1), 1–16 (2020)
4. Xueyan, Z., Guanhua, H.: Residents' exposure to indoor PM2.5 and its health risk assessment in rural dwellings of Liaoning Province. Build. Sci. **12**, 86–93 (2023)
5. Xueyan, Z., Yiming, Y., et al.: Diffusion characteristics of PM2.5 in rural dwelling under different daily life behavior a case study in rural Shenyang of China. Buiildings **43**(1), 1–18 (2021)
6. Chengjun, L., Hanqing, W., et al.: Diffusion characteristics of the industrial submicron particle under Brownian motion and turbulent diffusion. Indoor Built Environ. 1–14 (2022)
7. Xiujuan, W.: Study on Thermophoresis Effect of Indoor Fine Particles Hunan University of Technology, Hunan, China, 2016
8. Lu, W., Howarth, A.T., Jeary, A.P.: Prediction of airflow and temperature field in a room with convective heat source. Build. Environ. **32**(6), 541–550 (1997)
9. Emmerich, S.J., McGrattan, K.B., Kato, S., et al.: Application of a large eddy simulation model to study room airflow. ASHRAE Trans. **104**, 1128 (1998)
10. He, C., Ahmadi, G.: Particle deposition with thermophoresis in laminar and turbulent duct flows. Aerosol Sci. Technol. **29**(6), 525–546 (1998)
11. McLaughlin, J.B.: Aerosol particle deposition in numerically simulated channel flow. Phys. Fluids A: Fluid Dyn. (1989–1993) **1**(7), 1211–1224 (1989)
12. Yue, X.M.: Contrastive Analysis of Particle Distribution between Floor Heating and Radiator Heating Room, Taiyuan University of Technology, Dalian, China, 2010
13. Weizhuang, C., Weixue, C., Ru, C.: The research of the particulate matter in the radiant heating room. CC&AC **03**, 26–28 (2013)
14. Qinghua, C., Jiale, W., Dongsheng, L., et al.: Uniformity of temperature field in forced convection oven. Sci. Technol. Eng. **23**(22), 9509–9516 (2023)

Designing a Comprehensive Air Quality Monitoring Device: Electronics, Aesthetics, and Preliminary Data in Phu My Hung, Ho Chi Minh City

Mai Nguyen[1], Vy Luong[1], Than Diep[1,2], Phu Nguyen[1,3], Minh Duong[1,4], and Minh Doan[1](✉)

[1] Human-Centered Engineering, Fulbright University Vietnam, District 7, Ho Chi Minh City, Vietnam
minh.doan@fulbright.edu.vn

[2] Department of Mechatronics, Ho Chi Minh City University of Technology, Ho Chi Minh City, Vietnam

[3] Bosch Global Software Technologies Vietnam, Ho Chi Minh City, Vietnam

[4] Leon Legion Company Limited, Ho Chi Minh City, Vietnam

Abstract. This paper presents the design and implementation of a portable air quality monitoring device optimized for real-time data collection and transmission. Centered around the ESP32S3 micro-controller for its robust Wi-Fi capability, the device integrates two key sensor modules: the M701SC for measuring temperature, humidity, PM10, and PM2.5, and the SEN0377 for measuring CO and NO2 concentrations. To enhance its functionality, a NEO-7N module is employed for GPS-based location tracking, while an SD card module allows for onboard data logging. The device is powered by eight 21700 battery cells, ensuring continuous 24-h operation, including data acquisition and streaming. The enclosure, designed with aesthetics and durability, is 3D printed using PETG material. Temporal and spatial air quality data collected from various locations in the Phu My Hung area of Ho Chi Minh City demonstrate the device's effectiveness in real-world applications. Thiswork underscores the potential of integrating advanced electronics and industrial design to create efficient environmental monitoring solutions.

Keywords: Air quality monitoring · Internet of things · ESP32

1 Introduction

Over the past decade, air quality has captured increasing attention from governmental authorities, private sectors, and academic researchers. The growing recognition of air pollution's role in exacerbating public health crises, including respiratory and cardiovascular diseases [1–3], underscores the urgency of continuous monitoring. Environmental impacts further highlight the necessity of accurate air quality data [4, 5]. Effective monitoring enables the development of targeted policies and interventions while providing

C. Yuan (Ed.): ICRER 2024, ESE, pp. 144–154, 2025.
https://doi.org/10.1007/978-3-031-93295-3_12

communities with the information needed to safeguard their health. In this context, the advancement of air quality monitoring technologies has become a critical focus for ensuring a healthier and more sustainable future.

Air quality monitoring technology ranges from low-cost sensors, which provide accessible yet less precise measurements, to advanced, high-cost monitoring stations that offer highly accurate and comprehensive data. While the inexpensive sensors allow for widespread deployment and community-driven initiatives [6, 7], the expensive systems are often used by governmental agencies for regulatory purposes and detailed environmental assessments [8–10]. Both types of technology play complementary roles in understanding and managing air quality.

The primary objective of this work was to develop an affordable and efficient air quality monitoring system capable of accurately measuring critical environmental parameters, including PM2.5, PM10, CO, NO2 concentrations, temperature, and humidity. According to the U.S. Environmental Protection Agency (EPA), key air pollutants of concern include ozone, particulate matter (with diameters of 2.5 μ m and 10 μ m), carbon monoxide, lead, sulfur dioxide, and nitrogen dioxide [11]. Excluding lead, the EPA has published technical guidelines detailing the appropriate reporting of the Air Quality Index (AQI) for the remaining five pollutants, as regulated by the U.S. Clean Air Act [12]. These guidelines stipulate that a full 24-h data acquisition period is necessary to accurately calculate AQI values. Moreover, there is evidence indicating that the detection and measurement of ozone and sulfur dioxide can be influenced by NO2 and CO concentrations [13, 14], potentially complicating the simultaneous accurate measurement of all five pollutants without complex sensor calibration procedures.

In response to these challenges, the initial prototype was designed to report the AQI for three pollutants, PM2.5, PM10, and NO2, while maintaining continuous operation for over 24 h without requiring a constant power supply. Additionally, the system integrates GPS functionality to accurately record spatial data and ensures real-time data upload to an online database via a Wi-Fi connection. This combination of features provides a comprehensive and accessible solution for monitoring air quality across diverse environments. The present paper focuses on the hardware design of the system, while the corresponding software implementation is detailed in a companion paper presented within this conference series.

2 Methodology

A systematic approach was employed in the design and implementation of the air quality monitoring system. This section details the methodology adopted, including the selection and integration of electronic components, the design of the device enclosure, and the process of data acquisition and management.

2.1 Electronics

An overview of the system and relevant data flow can be seen in Fig. 1. The Espressif Systems ESP32S3 microcontroller was selected as the brain of the system for its compact size, reliable Wi-Fi capability, and number of general purpose input/output (GPIO) pins with different communication protocols available. The microcontroller unit (MCU) obtains data from 4 modules, namely M701SC, SEN0377, NEO-7N, and the touchscreen user input, while sending data to the SD card module and 2.8″ TFT LCD display.

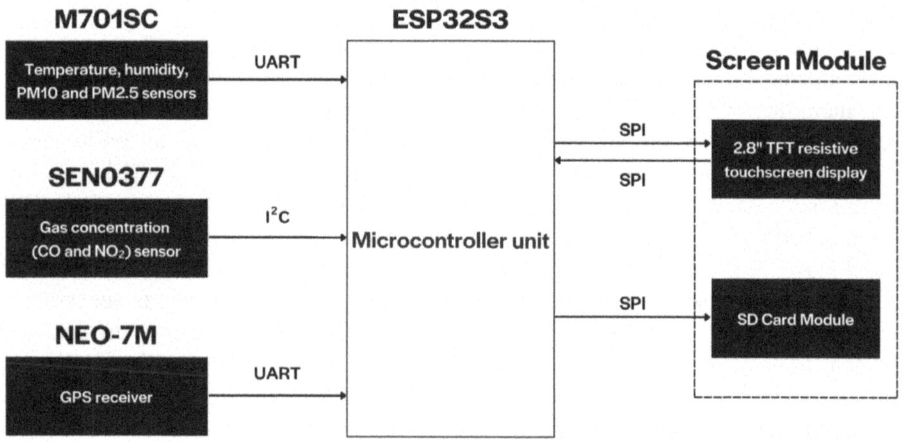

Fig. 1. An overview of the system and relevant data flow.

The M701SC module was selected for its comprehensive suite of sensors, including those for PM2.5, PM10, temperature, and humidity, which are essential for our air quality monitoring system. Its compatibility with the ESP32S3 microcontroller, facilitated by straightforward UART communication, made it an ideal choice for integration into our design. The particulate matter sensors in the M701SC module operate on the principle of laser scattering, detecting the concentration of airborne particles by measuring the light scattered by particles passing through a laser beam. Meanwhile, the temperature and humidity sensors rely on capacitive sensing technology, where changes in capacitance due to environmental conditions are converted into accurate temperature and humidity readings.

The SEN0377 module was chosen for its specialized sensors capable of detecting CO and NO2.The module communicate with the MCU via I2C protocol. The CO and NO2 sensors in the SEN0377 operate using electrochemical sensing technology, where the target gases react with an electrode, generating a measurable electrical current proportional to the gas concentration.

The NEO-7N GPS receiver was selected for its high precision in location tracking, which is essential for correlating air quality data with specific geographical locations. The module uses a combination of satellite signals to determine the device's position, providing accurate and real-time spatial data. The NEO-7N is also compatible with the MCU via the UART protocol. The NEO-7N uses a combination of satellite signals to determine the device's position, providing accurate and real-time spatial data critical for environmental monitoring applications.

In the presented system, the GPS receiver was supported by a 38 dBi active ceramic antenna through SMA connection.

The 2.8" TFT touchscreen module was integrated into the system for both display and interactive capabilities, containing the ILI9341 display driver and the XPT2046 touch controller. The module also includes an onboard SD card slot, providing an additional data storage option. All components communicate with the MCU through the SPI protocol, with the display and SD card functioning as output devices and the touch controller serving as an input. Although the ESP32S3 and most of the modules in the system operate at 3.3V, the M701SC module requires a 5V power supply, as outlined in Table 1. To ensure compatibility and protect the MCU's GPIO pins, a 4-channel level shifter was integrated into the system to step down the 5V UART signals to 3.3V. Consequently, two power rails are necessary to support both the M701SC module and the level shifter.

Table 1. Summary of data communications between the system modules and the microcontroller unit, their operating voltages, and power supply voltage.

Module/Chip	Operating Voltage (V)	Power Supply (V)	Communication Protocol
M701SC	5.0	5.0	UART
SENS0377	3.3	3.3–5.0	I^2C
NEO-7N	3.3	3.3–5.0	UART
ILI9341	3.3	3.3–5.0	SPI
XPT2046	3.3	3.3–5.0	SPI
SD card	3.3	3.3–5.0	SPI

Given the need for both 5V and 3.3V power rails, careful consideration was given to the power supply design. Lithium-ion (Li-ion) battery cells were selected over Li-polymer cells due to its higher energy density, longer lifespan, and better thermal stability, making it well-suited for ensuring the reliable operation of the device over extended periods. As Table. 1 suggests that all sensor modules whether with its own onboard 3.3V regulator or not can be powered at 5V, the main power supply was decided to be eight 21700 Li-ion batter cells in 1s8p configuration. The power consumption of each component was estimated based on the manufacturer's datasheet and is summarized in Table 2. The total power requirement was calculated to be 3.1W. Each 21700 cell, according to the manufacturer Lishen Battery [15], has a capacity of 4 Ah at a nominal voltage of 3.6V. To ensure at least 24 h of continuous operation without recharging, a minimum of five 4 Ah cells is required, providing approximately 27.9 theoretical hours of operation at a consumption rate of 3.1W. Since all of the published power consumption was at room temperature and the system was expected to operate at approximately 40°C, eight cells were implemented into the system for a margin of safety creating a theoretical of 37.2 h of operation. All 8 cells in parallel requires boosting the battery output voltage of 3.6V to 5V while reducing the need for battery balancing.

Table 2. Summary of the peak operating current and calculated power consumption at the rated voltage of all the system components.

Component	Peak Current (A)	Rated Voltage (V)	Power Consumption (W)
ESP32S3	0.38	3.3	1.25
M701SC	0.08	5.0	0.40
SEN0377	0.09	5.0	0.45
NEO-7N	0.07	3.0	0.21
SD module	0.10	3.3	0.33
TFT touchscreen	0.09	5.0	0.45

In actual implementation, a charging circuit board based on the FM5324GA chip was used for battery charging as well as voltage boosting with 2A maximum output current. The ESP32S3, M701SC, SEN0377, and TFT touchscreen including the onboard SD module were powered at this 5V boosted voltage. According to the manufacturer UBlox, the UBX-G7020-KT chip onboard of the NEO-7N module has a digital input/output voltage level of 1.65V–3.6V [16]. Therefore, the power supply to the GPS receiver was kept at 3.3V to protect the MCU's GPIO pins. The 3.3V power rail was taken from the ESP32S3's regulator. The system was, in practice, able to operate continuously for approximately 30 h, with all sensors active and real-time data being uploaded to Firebase via a Wi-Fi connection, before the battery voltage dropped below 2.8V and the over-discharge protection was activated.

2.2 Enclosure Design

The enclosure design presented in this paper represents the third iteration of the device development which can be seen in Fig. 2. The first iteration served as a proof-of-concept, utilizing inexpensive sensors housed in a 3D-printed PLA case. The second iteration incorporated the full suite of sensors detailed in this paper, enclosed within a simple wooden cuboid structure. The current, third iteration features a 3D-printed PETG enclosure, designed with comprehensive industrial design considerations to enhance both functionality and aesthetics.

In the second iteration, each side of the cuboid enclosure was dedicated to specific components. The LCD display was logically mounted on the front surface, while the battery cells were secured to the bottom surface for stability. The main circuit board, including power rails and signal lines, was positioned on the back surface, which also housed the main power button. The SEN0377 module was mounted on the left side, and the M701SC module was placed on the top side, which also contained the GPS antenna. The right side was reserved for the NEO-7N GPS receiver and the charger circuit.

Building upon this foundational layout, the third iteration, illustrated in Fig. 3, aimed to address both ergonomic and aesthetic considerations while enhancing the overall functionality of the device. The focus was on refining the user experience and improving the assembly process, which led to three key design advancements. First, the front surface

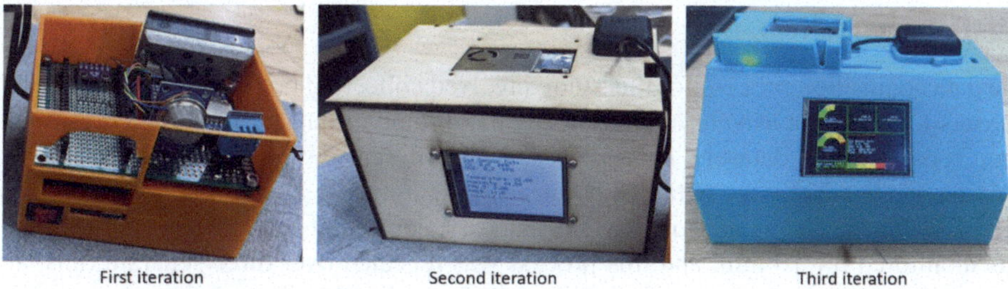

Fig. 2. The device enclosure development through 3 iterations.

was split into a vertical section and an angled section at 45°, specifically designed to improve user ergonomics. This angle was chosen to make it easier for users to optimize viewing comfort for users, as the device is often placed below eye level. Second, the enclosure's construction was refined using 3D printing technology, allowing for a seamless assembly of the six surfaces, along with custom containers for the M701SC and SEN0377 modules, enhancing both functionality and visual appeal. Fitted inserts were employed to connect the surfaces, and brass nuts were embedded into designated holes, enabling the secure mounting of modules with standard bolts while maintaining a smooth exterior without visible fasteners. Third, the graphical user interface (GUI) was significantly improved (Fig. 4), allowing users to easily switch Wi-Fi connections, gather real-time data, and visualize the information on-screen. The details of the GUI implementation will be thoroughly discussed in the companion paper.

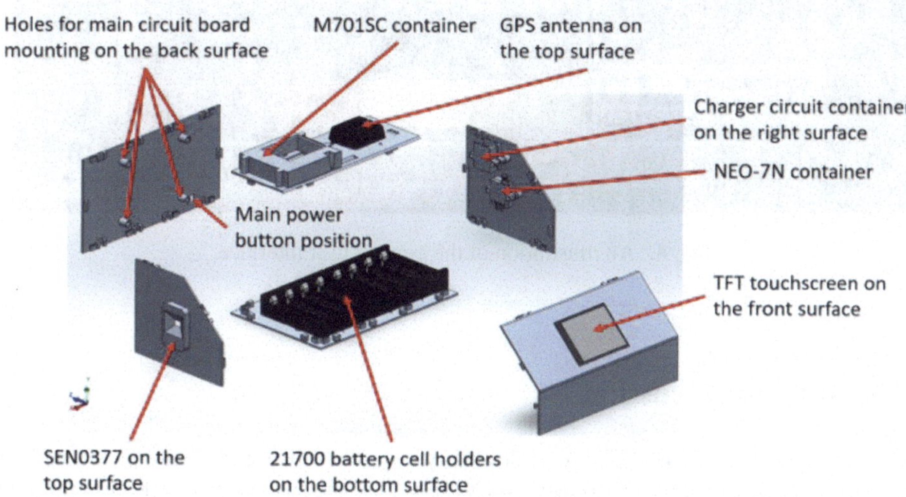

Fig. 3. The exploded view the CAD design.

2.3 Data Acquisition

The data acquisition process for this study involved collecting both temporal and spatial datasets to comprehensively assess air quality. Temporal data were gathered through three separate 24-h datasets, each recorded on different days at a fixed location. In addition, spatial data were obtained by measuring air quality at seven distinct locations within the Phu My Hung area of Ho Chi Minh City. For spatial consistency, each location was monitored for 10 min, and this process was repeated over three separate runs, all conducted within the same time window of 4:00 to 5:30 PM.

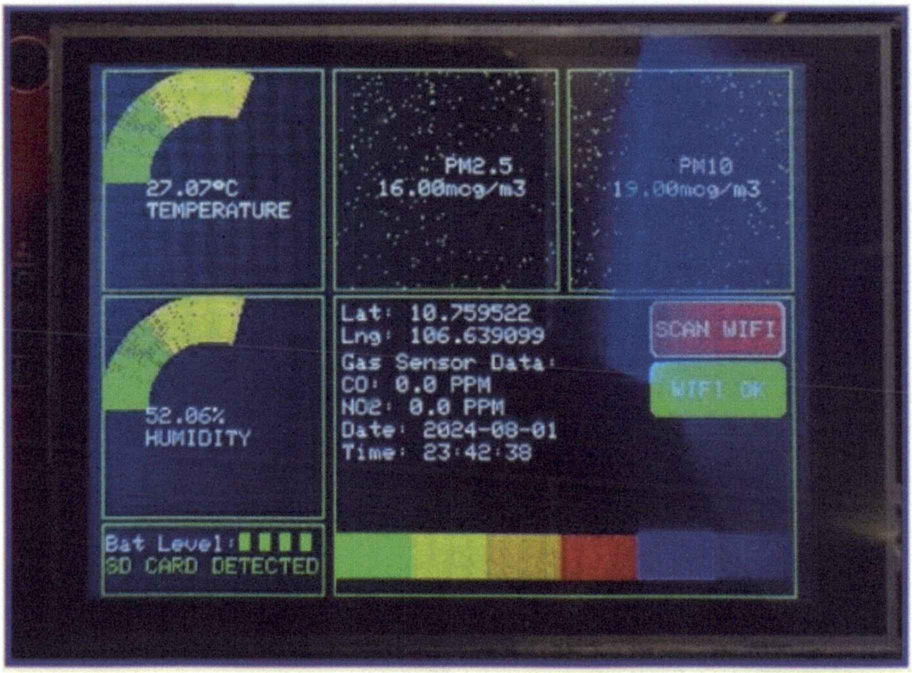

Fig. 4. An illustration of the graphic user interface.

3 Result and Discussion

First and foremost, the SEN0377 module ability to detect CO and NO2 in typical ambient air conditions was evaluated. To test the CO sensor's responsiveness, carbon monoxide was introduced using a pocket lighter, and observed a rapid response within just a few seconds, as illustrated in Fig. 5. For NO2, concentrations were occasionally measured to be around 0.2 ppm, consistent with findings by Hesterberg et al. [17]. These low levels reflect typical urban background concentrations, confirming the module's ability to detect NO2 when present.

A 24-h continuous measurement of PM10, PM2.5, temperature, and humidity was conducted at a fixed location with coordinates of 10.73644 latitude and 106.741938 longitude, as illustrated in Fig. 6. The data, recorded from 9:00 am to 9:00 am the following day, revealed expected trends: temperature increased during the morning while humidity decreased, and the reverse occurred during the night. Interestingly, there was evidence of a negative correlation between particulate matter concentrations and temperature, and a positive correlation with humidity. Particulate matter concentrations peaked around 7:00 pm, suggesting potential environmental or human activity influences. However, additional data is needed to confirm the consistency of these trends across different days and conditions.

Preliminary spatial data on particulate matter concentrations, specifically PM10 and PM2.5, was obtained as shown in Fig. 7. The yellow-circled point represents a commercial area, while the green-circled point corresponds to an inactive construction site. The remaining points are located in residential areas, all within Phu My Hung and its surrounding vicinity. Interestingly, despite the presence of construction materials, the inactive construction site exhibited lower levels of both PM10 and PM2.5 than some residential locations. These measurements were collected between 4:00 pm and 5:30 pm, indicating potential variations in particulate matter distribution based on localized factors. Further investigation with higher resolution and simultaneous data acquisition will be needed to fully understand these spatial discrepancies.

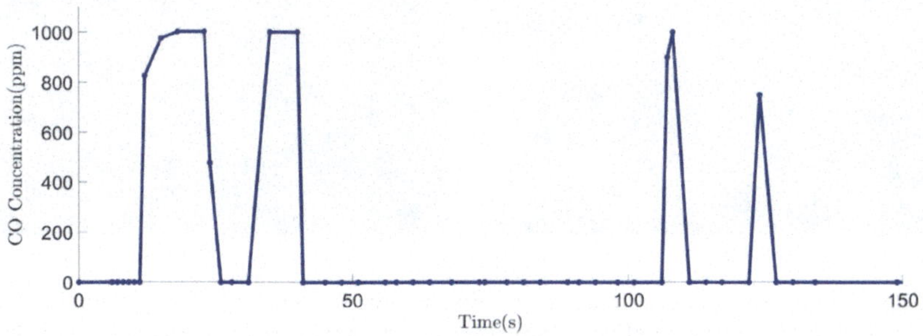

Fig. 5. Of the SEN0377 module's responsiveness to carbon monoxide in ambient air.

Lastly, it was observed that the NEO-7N GPS module occasionally required more than 20 min to acquire a signal and return accurate coordinates. This delay created challenges, especially during field measurements where rapid location tracking was necessary. To mitigate this, GPS data were cross-checked with mobile phone readings to ensure accuracy. In some instances, the displayed coordinates were directly obtained from the phone's GPS to provide timely source for spatial data.

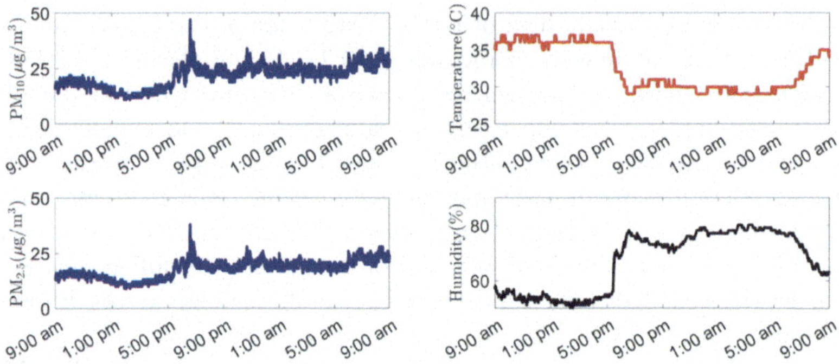

Fig. 6. An example of temperature, humidity, and particulate matter concentration data measured for 24 h continuously.

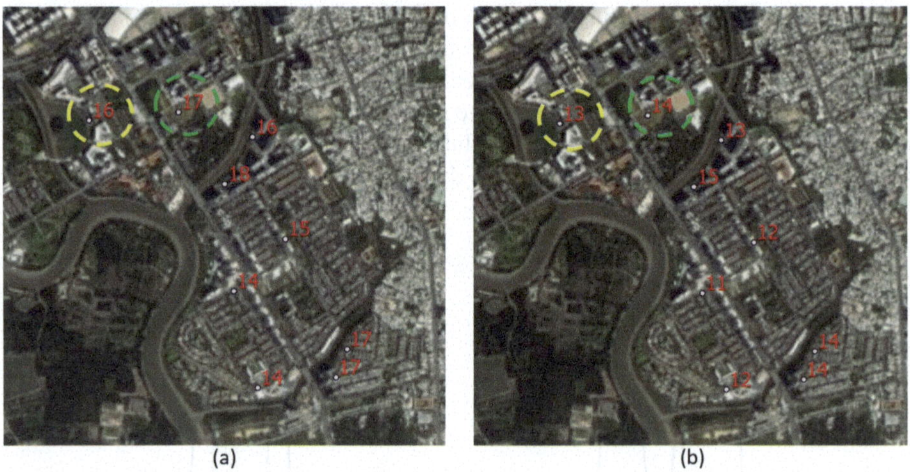

Fig. 7. Preliminary spatial particulate matter concentration data: (a)PM10 and (b)PM2.5. The point enclosed in the yellow circle is a commercial area while the point in green is an inactive construction spot. The remaining points are located in residential areas. The concentration values were 10-min averaged at each spot.

4 Conclusion

This paper presented the design and implementation of a portable, cost-effective air quality monitoring system that effectively measures critical environmental parameters, including PM2.5, PM10, CO, NO2 concentrations, temperature, and humidity. The device, centered around the ESP32S3 microcontroller, was engineered with a focus on continuous 24-h operation, real-time data transmission, and ease of use, incorporating a 45' angled display for optimal user interaction. The modular design, facilitated by 3D printing technology, allowed for a seamless assembly with hidden fasteners, contributing to both the aesthetic and functional appeal of the system. The device's ability to log data on an SD card, upload it to an online database, and track precise GPS locations demonstrated its versatility and effectiveness in various urban environments. The spatial and temporal data collected in Phu My Hung, Ho Chi Minh City, confirmed the system's reliability in real-world applications. This work lays the groundwork for further advancements in environmental monitoring technology, with future improvements focusing on enhancing the software side of the system, as discussed in a companion paper.

The spatial and temporal data collected in Phu My Hung, Ho Chi Minh City, confirmed the system's reliability in real-world applications. Future work will focus on scaling the system by manufacturing multiple devices to measure air quality at different locations simultaneously. To facilitate this, a custom PCB will be designed to streamline the manufacturing process. Additionally, the device could be adapted for indoor air quality monitoring, which may allow for a more compact and lighter design by removing the GPS module and reducing the number of battery cells required.

References

1. Lee, B.J., Lee, B., Lee, K.: Air pollution exposure and cardiovascular disease. Toxicol. Res. **30**, 71–75 (2014)
2. Mirabelli, M.C., et al.: Air quality awareness among U.S. adults with respiratory and heart disease. Am. J. Prev. Med. **54**(5), 679–687 (2018)
3. Shahriyari, H.A., et al.: Air pollution and human health risks: mechanisms and clinical manifestations of cardiovascular and respiratory diseases. Toxin Rev. **41**(2), 606–617 (2021)
4. Fiore, A.M., Naik, V., Leibensperger, E.M.: Air quality and climate connections. J. Air Waste Manag. Assoc. **65**, 645–685 (2015)
5. von Schneidemesser, E., et al.: Chemistry and the linkages between air quality and climate change. Chem. Rev. **115**, 3856–3897 (2015)
6. Cheng, Y., et al.: Aircloud: a cloud-based air-quality monitoring system for everyone. In: Proceedings of the 12th ACM Conference Embedded Network Sensor System, pp. 251–265. SenSys'14, Association for Computing Machinery, New York, NY, USA (2014)
7. Kumar, S., Jasuja, A.: Air quality monitoring system based on IoT using raspberry pi. In: International Conference on Computing, Communication and Automation (ICCCA), pp. 1341–1346 (2017)
8. Liu, J.H., et al.: Developed urban air quality monitoring system based on wireless sensor networks. In: 5th International Conference on Sensing Technology, pp. 549–554 (2011)
9. Mansour, S., Nasser, N., Karim, L., Ali, A.: Wireless sensor network-based air quality monitoring system. In: International Conference on Computing, Networking and Communications (ICNC), pp. 545–550 (2014)
10. Zheng, K., Zhao, S., Yang, Z., Xiong, X., Xiang, W.: Design and implementation of LPWA-based air quality monitoring system. IEEE Access **4**, 3238–3245 (2016)
11. United States Environmental Protection Agency: Criteria air pollutant information (2024)
12. United States Environmental Protection Agency: Technical assistance document for the reporting of daily air quality – the air quality index (AQI) (2024)
13. Markus, L., Marc, B., Andreas, S.: Ozone detection in the ppb-range with improved stability and reduced cross sensitivity. Sens. Actuators B: Chem. 130(1), 367–373 (2008), 2008 Proceedings of the Eleventh International Meeting on Chemical Sensors IMCS-11
14. Qi, H., Zhao, X., Xu, Y., Yang, L., Liu, J., Chen, K.: Rapid photoacoustic exhaust gas analyzer for simultaneous measurement of nitrogen dioxide and sulfur dioxide. Anal. Chem. **96**(13), 5258–5264 (2024)

15. Lishen: 21700 power cylindrical battery LR2170LA (2024)
16. UBlox: UBX-G7020: u-blox 7 GNSS chips (2024)
17. Hesterberg, T.W., Bunn, W.B., McClellan, R.O., Hamade, A.K., Long, C.M., Valberg, P.A.: Critical review of the human data on short-term nitrogen dioxide (NO_2) exposures: evidence for NO_2 no-effect levels. Crit. Rev. Toxicol.Toxicol. **39**(9), 743–781 (2009)

Development of Geopolymer Mortar from Palm Oil Boiler Ash and Alumina Powder: A Process Study

Eugene Jian Jun Chai[1(✉)], Henry Chee Yew Foo[1], Ziad Bennour[1], Hisham Khaled Ben Mahmud[2], and Shiew Wei Lau[1]

[1] Department of Chemical and Energy Engineering, Curtin University Malaysia, CDT 250, 98009 Miri, Malaysia
Eugenechai@postgrad.curtin.edu.my

[2] Department of Petroleum Engineering, Universiti Teknologi PETRONAS, Persiaran UTP, 32610 Seri Iskandar, Perak , Malaysia

Abstract. Geopolymer is a promising green building material compared to Portland cement. It is produced by activating the aluminosilicate materials rich in Si and Al with alkaline solution (sodium hydroxide and sodium silicate) (NaOH and Na_2SiO_3). As one of the largest palm oil producers globally, Malaysia generates significant amounts of waste, including palm oil fuel ash (POFA) and palm oil boiler ash (POBA), during the processing of palm oil. While POBA is rich in SiO_2, it lacks sufficient Al_2O_3 to undergo geopolymerization, limiting its potential for geopolymer production. This experiment studies the effect of NaOH concentration (NaOH conc), solid-to-liquid mass ratio (S/L mass ratio), and percentage mass of alumina powder (Al_2O_3) replacement on the compressive strength of POBA-based geopolymer. The Na_2SiO_3/NaOH volume ratio and curing temperature were fixed at 2 and 80 °C, respectively. During the experiment, a total of 12 designs were evaluated to optimize the parameters. The study ranges identified for NaOH concentrations, B/L mass ratio, and percentage mass of Al_2O_3 for POBA-based geopolymer were 5 M to 20 M, 1 to 2.5, and 5% to 30%, respectively. The experiment result indicates that the optimum mixing ratio for POBA-based geopolymer has NaOH concentration of 15 M, S/L mass ratio of 1.5, and 25% Al_2O_3 replacement, yielding the highest compressive strength at 7 days.

Keywords: Geopolymer · Palm oil boiler ash (POBA) · Compressive strength

1 Introduction

The Ordinary Portland Cement (OPC) is often used for construction and oil well cement. However, the cement industry has contributed to the emission of carbon dioxide (CO_2), which leads to global climate change [1]. The OPC production is about 1.5 billion tons annually, with an average of 6% of total emission from other sectors around the world [2]. Therefore, developing a greener cement is a prominent substitute for OPC. Geopolymer has been found for potential candidates to substitute OPC as it has lower

C. Yuan (Ed.): ICRER 2024, ESE, pp. 155–169, 2025.
https://doi.org/10.1007/978-3-031-93295-3_13

environmental impact and better properties compared to OPC [3]. The raw materials of geopolymer are aluminosilicate-rich materials such as fly ash, metakaolin, slag, and various agricultural byproducts [4, 5]. Upon contact with alkaline solution, the raw material undergoes geopolymerization process that dissolves into aluminum (Al) and silicon (Si) and forms sodium aluminosilicate hydrate (N-A-S-H) gel as the final product (Fig. 1). As an agricultural country, Malaysia is rich in palm oil, which produces large amounts of palm oil ash that can be used as raw material for geopolymer [6]. However, the palm oil boiler ash (POBA) lacks Al_2O_3, which cannot undergo geopolymerization process completely [7]. In order for the geopolymer to solidify, the presence of Al is essential to initiating the formation of the Al-O-Si bond [8]. In this study, the effect of NaOH concentration, S/L mass ratio, and percentage of Al_2O_3 replacement on the compressive strength of the geopolymer after curing for 7 days is investigated. Based on the result, the best ratio for the mixing of POBA-based geopolymer can be derived.

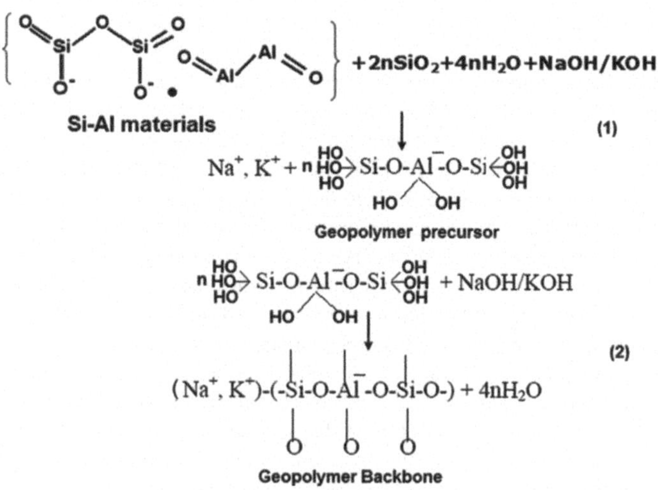

Fig. 1. Geopolymerization process [9].

2 Literature Review

The raw material of palm oil-based geopolymer can be sourced from palm oil fuel ash (POFA) or palm oil boiler ash (POBA). In fact, there is a slight difference between the chemical composition of POFA and POBA based on X-ray fluorescence (XRF), as shown in Table 1. Currently, there is intensive research on POFA-based geopolymer, while there is only a little research on POBA-based geopolymer [7]. Basically, POFA is mixed with other components such as rice husk ash (RHA) and metakaolin (MK) to achieve higher compressive strength. For instance, Salih et al. [10] investigated the effect of different combination ratios of Na_2SiO_3 and NaOH with POFA under 60 °C. They found that the Na_2SiO_3/NaOH ratio is directly proportional to the compressive strength of the geopolymer. The Na_2SiO_3/NaOH ratio of 2.5 and 3 yielded a compressive strength of 32 MPa after 28 days. Mijarsh et al. [11] used the Taguchi method to optimize the amount of $Ca(OH)_2$, $Al(OH)_3$, and silica fume for the maximum compressive strength of treated POFA (TPOFA)-based geopolymer. They stated that the optimum amount of TPOFA is 65 wt%, with 20 wt% of $Ca(OH)_2$, 5 wt% of silica fume, and 10 wt% of $Al(OH)_3$. This results in a compressive strength of about 47.27 MPa in 7 days. Moreover, Isa and Awang [12] mixed POFA and ground granulated blast furnace slag (GGBS) with alkaline solution (NaOH and Na_2SiO_3) under ambient conditions. The inclusion of GGBS into POFA acts as a Ca and Al source for the formation of calcium aluminosilicate hydrate (C-A-S-H) gel, which strengthens the geopolymer. Moreover, the inclusion of 40% GGBS with POFA produces a final compressive strength of 28.17 MPa on 28 days. Furthermore, Ranjbar et al. [13] investigated the effect of POFA replacement on fly ash (FA)-based geopolymer. They found that the higher the SiO_2/Al_sO_3 ratio, the lower the early compressive strength. As Al tends to dissolve easily, therefore the Al would react in early stage and lead to the deficiency of Al in later stages, resulting in lower compressive strength. Hawa et al. [14] blended palm oil ash with MK under ambient temperature and 80 °C for 2 h. They indicated that the MK with 5% palm oil ash yielded the maximum compressive strength of 62 MPa curing under 80 °C. Alias Tudin et al. [15] investigated the properties of POFA and RHA-based geopolymers. They indicated that the RHA-based geopolymer has higher compressive strength compared to POFA-based geopolymer due to the higher Si content in RHA. This results in more Si-O-Si bonds in RHA- based geopolymers compared to POFA-based geopolymers. As the Si-O-Si bonds are stronger than Al-O-Si and Al-O-Al bonds, therefore leading to a higher overall compressive strength.

Table 1. XRF result of POFA and POBA [7, 10].

Composition	POFA (%)	POBA (%)
SiO_2	47.37	40.60
Al_2O_3	3.53	3.71
Fe_2O_3	6.19	15.74
CaO	11.83	19.60
MgO	4.19	1.30
P_2O_5	3.31	2.73
SO_3	1.22	0.44
Loss on ignition (L.O.I)	1.84	5.01

3 Materials and Methods

3.1 Materials

The raw materials required are POBA and alumina (Al_2O_3) powder. The POBA and alumina powder were purchased from Batu Pahat, Johor, Malaysia. Moreover, the NaOH and Na_2SiO_3 were purchased online in Malaysia.

3.2 Geopolymer Synthesis

Initially, the POBA was dried in oven under 100 °C for 24 h to remove the moisture. Then, the POBA was grounded in ball mill to obtain a smaller particle size. The grounded POBA was sieved through a 75 μ m sieve. The alkaline solution used is a combination of Na_2SiO_3 and NaOH with a constant volume ratio of 2:1. Moreover, the NaOH was dissolved in distilled water to obtain the desired concentration. They were premixed 24 h before usage. The synthesis of POBA-based geopolymer was done by mixing alkaline solution, POBA (Fig. 2), and alumina powder based on Table 2.

Fig. 2. Palm Oil Boiler Ash (POBA)

Table 2. Design of Experiment

Case	Run	Na_2SiO_3/NaOH volume ratio	NaOH conc (M)	Al_2O_3 (%)	S/L mass ratio (g/g)
1	1	2	15	0	1
	2				1.5
	3				2
	4				2.5
Case	Run	Na_2SiO_3/NaOH volume ratio	S/L mass ratio (g/g)	Al_2O_3 (%)	NaOH conc (M)
2	5	2	1.5	0	5
	6				8
	7				11.5
	8				20
Case	Run	Na_2SiO_3/NaOH volume ratio	NaOH conc (M)	S/L mass ratio (g/g)	Al_2O_3 (%)
3	9	2	15	1.5	5
	10				10
	11				25
	12				30

First, POBA and alumina powder were mixed manually for 5 min to obtain a homogeneous mixture. Then, the mixture was poured into alkaline solution gradually. After that, it was mixed using Fann constant speed mixer with a speed of 12,000 rpm for 5 min. Then, the geopolymer slurries were cast in cylinder silica mold with a height and diameter of 50 mm. After that, the geopolymer slurries were cured in oven under 80 °C for 24 h. The silica mold must be covered with cling film to prevent moisture loss during the curing of the geopolymer. After 24 h of curing, the geopolymer was removed from oven and left at ambient temperature to further cure for 7 days. After 7 days, it was subjected to uniaxial compressive strength (UCS) test and Fourier-transform infrared (FTIR) spectroscopy to examine the properties of the geopolymer. The overall experimental procedure was summarized in Fig. 3.

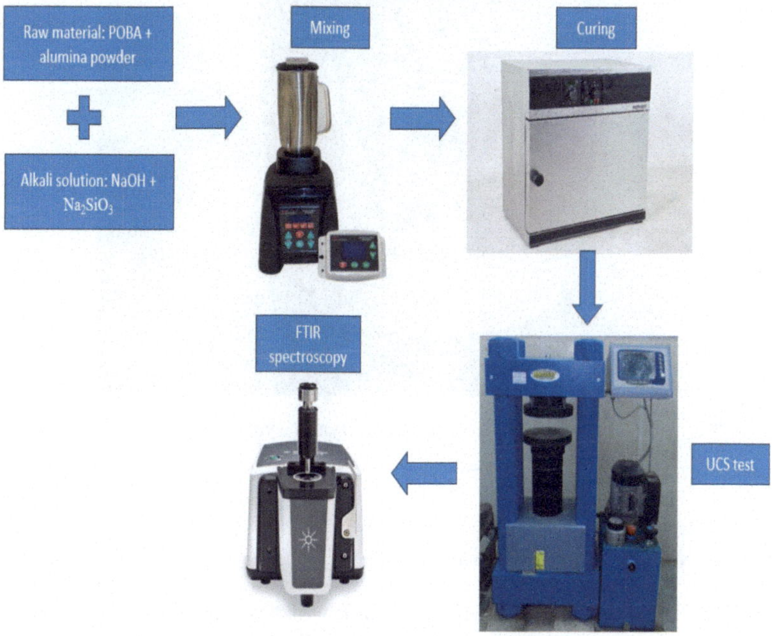

Fig. 3. Overall experimental procedure.

3.3 Design of Experiment

The parameters studied on the POBA-based geopolymer were divided into 3 cases (Table 2). Case 1 and 2 investigated the compressive strength of geopolymer by varying the solid-to-liquid mass ratio (S/L) and NaOH concentration without the addition of Al_2O_3. On the other hand, case 3 investigated the compressive strength of geopolymer by varying the percentage replacement of Al_2O_3. The experiment data were analyzed by using OriginLab software. From the experiment result, the optimum mixing ratio for POBA-based geopolymer can be determined by obtaining the ratio from case 1, 2, and 3, which yield the highest compressive strength.

4 Results and Discussion

Figure 4 shows the POBA-based geopolymer after curing for 7 days. The appearance of POBA-based geopolymer is black with ceramic-like properties.

Fig. 4. POBA-based geopolymer synthesized.

4.1 Effect of NaOH Concentrations

The function of NaOH is to facilitate the dissolution of raw materials to undergo geopolymerization reactions [11]. In this experiment, the POBA-based geopolymer was cured with different concentrations of NaOH based on case 2 shown in Table 2. Moreover, the compressive strength of geopolymer with NaOH concentration from 5 M to 20 M was summarized in Table 3.

Table 3. Compressive strength of POBA-based geopolymer with varying NaOH conc.

NaOH concentration (M)	UCS (MPa)
5	3.82
8	7.8
11.5	9.56
15	10.76
20	10.5

Based on Fig. 5, it was observed that the compressive strength is directly proportional to NaOH concentration. As the concentration of NaOH increases, the Si and Al from raw materials increase, which enhances the formation of Al-O-Si and Si-O-Si bonds during the polycondensation [16]. This can be observed from the increase in temperature in the mixer due to the heat release from mixing POBA with alkaline solution. However, the compressive strength drops slightly from 10.76 MPa to 10.5 MPa as the NaOH concentration increases from 15 M to 20 M, indicating the optimum NaOH concentration is 15 M.

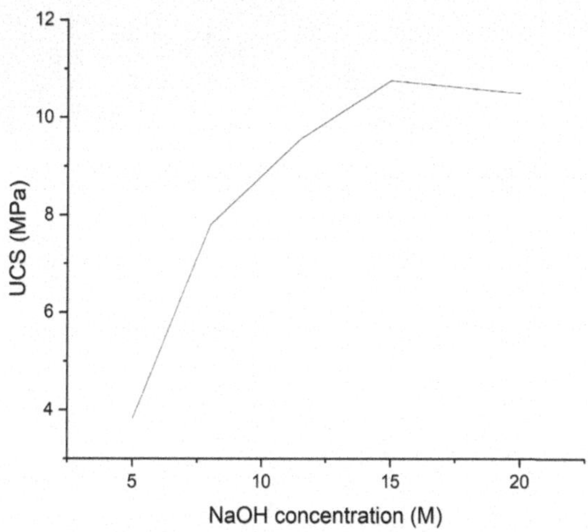

Fig. 5. Compressive strength of POBA-based geopolymer vs NaOH concentration.

On the other hand, it was also observed that excess NaOH concentration would lead to the precipitation of sodium carbonate (Na_2CO_3) on the surface of geopolymer, which is known as efflorescence effect. This is due to the reaction of CO_2 with the excess sodium ions [17]. Figure 6 shows the effect of efflorescence due to the usage of 15 M NaOH (left side).

4.2 Effect of S/L Mass Ratio

The B/L mass ratio controls the workability of geopolymer slurry. The geopolymer in this experiment was mixed with a varying S/L mass ratio based on case 1 as shown in Table 2. Furthermore, the compressive strength of geopolymer with S/L mass ratio from 1 to 2.5 was summarized in Table 4.

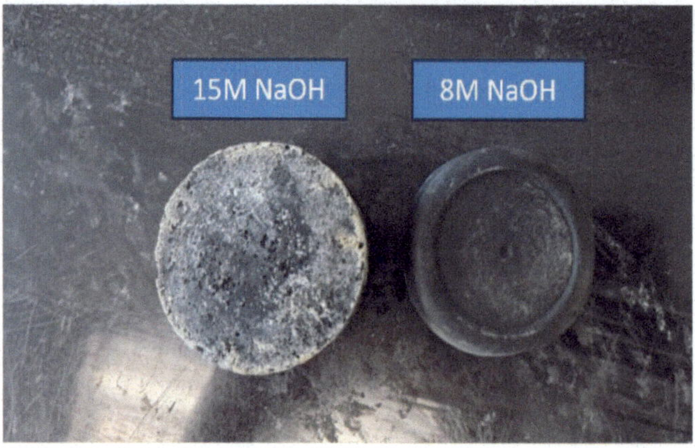

Fig. 6. Efflorescence effect on geopolymer.

Table 4. Compressive strength of POBA-based geopolymer with varying S/L mass ratio.

S/L mass ratio	UCS (MPa)
1	5.23
1.5	10.76
2	10
2.5	8

A higher S/L mass ratio indicates more POBA has been added into the geopolymer during the mixing stage. This would lead to the low workability of geopolymer slurry, resulting in slurry with high viscosity and difficulty in mixing. Moreover, the raw material in excess may also lead to the insufficient wetting of raw materials, leading to difficulty in molding [16]. The low water content during geopolymer mixing can potentially lead to a porous geopolymer matrix, reducing the overall compressive strength. This can be observed in Fig. 7, in which the compressive strength of geopolymer decreases from 10.76 MPa to 8 MPa as the S/L mass ratio increases from 1.5 to 2.5. On the other hand, a lower B/L mass ratio indicated more alkaline solution has been added into the geopolymer during the mixing stage. Although it improves the workability of geopolymer slurry, it reduces the compressive strength of the geopolymer as the water content increases. The excess water dilutes the polycondensation of geopolymer by leaching away Si and Al away from the reaction zone [11]. Therefore, finding the optimum S/L mass ratio is important during the mixing of geopolymers. From the experiment, it was found that the optimum S/L mass ratio is 1.5, yielding the highest compressive strength without compromising the workability of geopolymer slurry.

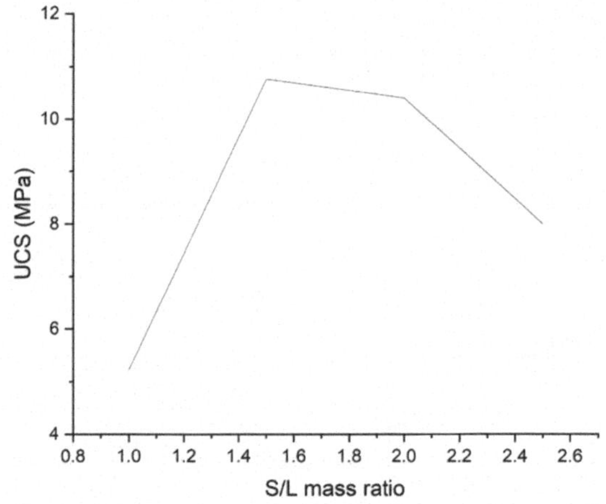

Fig. 7. Compressive strength of POBA-based geopolymer vs S/L mass ratio.

4.3 Effect of Al_2O_3 Replacement

The geopolymer in this experiment was mixed with a varying ratio of Al_2O_3 based on case 3 as shown in Table 2. The addition of Al_2O_3 into POBA accelerates the formation of geopolymer gel, thus leading to the hardening of geopolymer [9]. The Al_2O_3 decreased the setting time of geopolymer significantly. Also, the presence of Al-O-Si bonds was detected after the addition of Al_2O_3 [18]. On the other hand, the addition of Al_2O_3 leads to a decrease in the workability of the geopolymer [19]. This was also observed during the production of geopolymer as the geopolymer slurries become viscous after the addition of Al_2O_3. The compressive strength of POBA-based geopolymer after the addition of Al_2O_3 was summarized in Table 5.

Table 5. Compressive strength of POBA-based geopolymer with different Al_2O_3 replacement.

Al_2O_3 (%)	UCS (MPa)
5	13.55
10	15.14
25	18.55
30	18.3

Based on Fig. 8, the addition of Al_2O_3 from 5% mass to 25% mass increases the compressive strength from 13.55 MPa to 18.55 MPa. In summary, Al_2O_3 is important during the production of geopolymer as it promotes the geopolymerization process and strengthens the matrix of geopolymer.

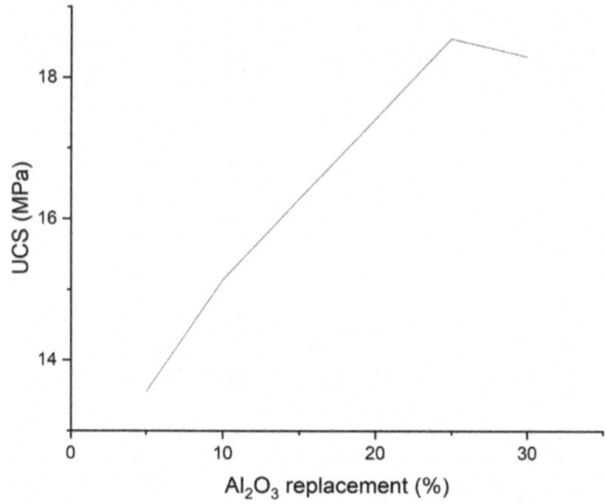

Fig. 8. Compressive strength of POBA-based geopolymer vs Al_2O_3 replacement.

5 Results Analysis

5.1 Uniaxial Compressive Strength (UCS)

Based on the experimental results, the optimum NaOH concentration, S/L mass ratio, and Al_2O_3 replacement are 15 M, 1.5, and 25%, which give a compressive strength of 18.55 MPa. As compared with POBA-based geopolymer without the addition of Al_2O_3 (run 2), the addition of Al_2O_3 into POBA-based geopolymer (run 11) has increased the compressive strength by 72% under the same mixing condition. Figure 9 shows the stress-strain curve for run 2 and run 11 of POBA-based geopolymer analyzed using OriginLab software. Moreover, the engineering properties of geopolymer such as Young's Modulus (E), fracture strength (σ_f), and toughness were calculated using OriginLab software. The values of E, σ_f, and toughness of POBA-based geopolymers were summarized in Table 6.

From Fig. 9, it was observed that the POBA-based geopolymer is brittle material for both cases, as the stress-strain curve is linear up to the fracture point. Moreover, E is a measure of the material's rigidity. It quantifies how much a material deforms under load, expressed as Eq. (1) [20]. Typical E of concrete and geopolymer are 18–40 GPa and 22.95–30.84 GPa [21, 22]. Based on Eq. (1), the Young's Modulus of POBA-based geopolymer can be calculated from the slope of the linear section of the stress-strain curve in Fig. 9. The calculated E for POBA-based geopolymer under run 11 was 8.34 GPa, which is lower than the typical value. This may be due to the incomplete hydration of geopolymer, as it needs 28 days of complete hydration. On the other hand, σ_f is the amount of strength a material can withstand before fracture. It is the highest point in the stress-strain curve, which corresponds to the compressive strength of POBA-based geopolymer. Lastly, toughness is the capacity of a material to absorb energy until fracture. It is the area of the stress-strain curve with a unit of energy per volume. Moreover, the material with high strength usually has high toughness.

$$E = \frac{\sigma}{\varepsilon} \tag{1}$$

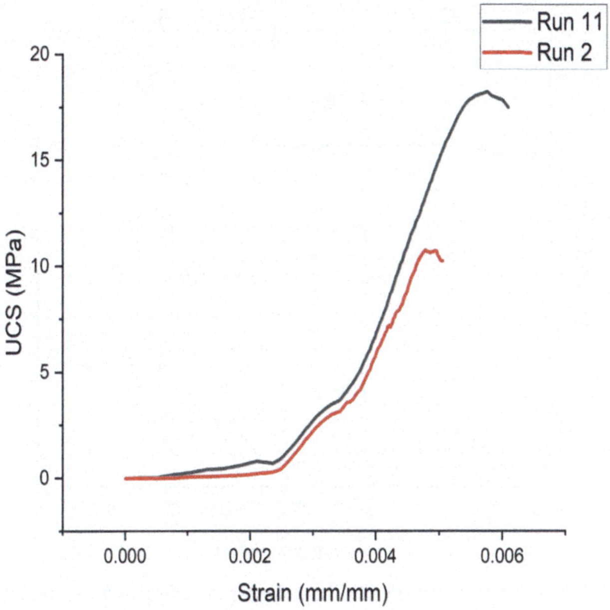

Fig. 9. Stress-strain curve for POBA-based geopolymer produced under run 2 and run 11 cured for 7 days.

Table 6. E, σ_y , and toughness of POBA-based geopolymer.

	Run 2	Run 11
Young's Modulus (E) (GPa)	2.43	8.34
Yield strength (σ_f) (MPa)	10.76	18.55
Toughness (energy/volume)	0.11	0.04

5.2 FTIR Spectroscopy

The FTIR test was carried out using Cary 630 FTIR spectrometer from Agilent. First, about 5 mg of sample was first ground into smaller particle sizes using mortar and pestle. Then, the sample was analyzed using attenuated total reflectance (ATR) mode. The wavelength obtained ranged from 400–4000 cm^{-1}. By comparing the intensity of each FTIR peak, the changes in the concentration of each phase can be quantified. Figure 10 shows the FTIR result for POBA, run 2, and run 11 of the POBA-based geopolymer. As the geopolymerization proceeds, the N-A-S-H and calcium silicate hydrate (C-S-H) gels increase. Moreover, Table 7 shows the summary of each wavelength and functional group associated in POBA-based geopolymer.

There are 2 wavelength regions that are used to indicate the degree of geopolymerization commonly. First, the peak from the Si-O-Si stretching vibration of quartz at 777–795 cm^{-1} decreases as the geopolymerization proceeds [13]. Then, the decreases in

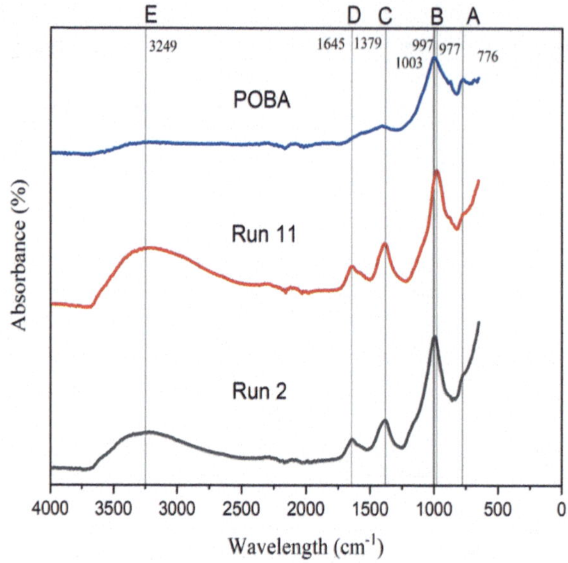

Fig. 10. FTIR result for POBA and POBA-based geopolymer produced under run 2 and 11.

Table 7. Summary of FTIR results.

Band	Wavelength (cm^{-1})	Functional Group
A	776	Si-O-Si stretching
B	1003	Si-O/Al-O stretching
E	3249	O-H stretching

Band	Wavelength (cm^{-1})	Functional Group
A	776	Si-O-Si stretching
B	977	Si-O/Al-O stretching
C	1379	CO_3^{2-} stretching
D	1645	O-H bending
E	3249	O-H stretching

Band	Wavelength (cm^{-1})	Functional Group
A	776	Si-O-Si stretching
B	997	Si-O/Al-O stretching
C	1379	CO_3^{2-} stretching
D	1645	O-H bending
E	3249	O-H stretching

peak from 1000 to 970 indicate the substitution of Si^{4+} by Al^{3+}, thus forming the Al-O-Si bond [23, 24]. This bond is responsible for the solidification of the geopolymer. From

Fig. 11, it was observed that the POBA-based geopolymers developed from POBA have a decrease in wavelengths from 1000 cm^{-1} to 991 cm^{-1} and 977 cm^{-1} respectively. This indicates the presence of geopolymerization and the formation of Al-O-Si bonds in the POBA-based geopolymer. Moreover, it was observed that the peak of band B under run 11 is higher than run 2 in Fig. 11. This indicates there are more N-A-S-H gels formed with the addition of Al_2O_3 to POBA-based geopolymer. As a result, the compressive strength of POBA-based geopolymer has increased after the addition of Al_2O_3. On the other hand, the presence of band C was due to the efflorescence effect [23]. As shown in Fig. 10, the peak in run 11 is higher than the peak in run 2, which means there is more Na_2CO_3 formed in the case of run 11. This can also be observed visually, as there is white solid of Na_2CO_3 precipitated on the surface of the POBA-based geopolymer in the case of run 11, as shown in Fig. 6. Then, the bands D and E correspond to the water absorbed inside the geopolymer. The water content in geopolymer is important, as low water content may result in geopolymer with high porosity due to insufficient wetting. As POBA was dried before mixing, the water content is the lowest, which can be observed from the intensity of peak E in Fig. 10.

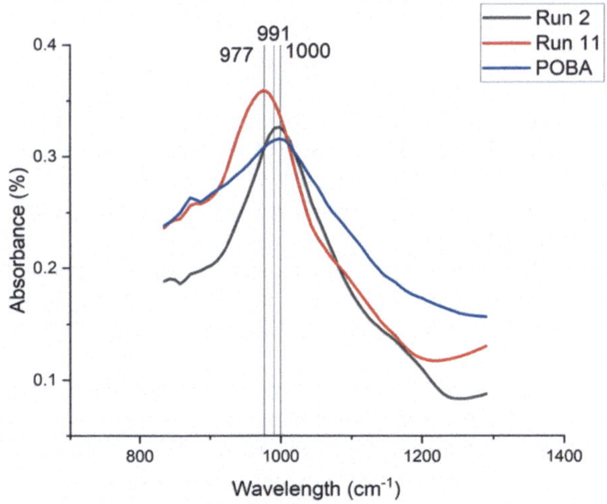

Fig. 11. FTIR result for POBA and POBA-based geopolymer produced under run 2 and 11 at wavelength region from 800 to 1300 cm^{-1}.

6 Conclusion

Utilizing POBA- based geopolymers instead of OPC offers significant environmental, economic, and performance advantages that align with SDG goals and help reduce CO_2 emissions. This experiment has shown that NaOH concentration, S/L mass ratio, and the addition of Al_2O_3 affect the compressive strength of POBA-based geopolymer. As compared with the POBA-based geopolymer without the addition of Al_2O_3, the addition of Al_2O_3 during the mixing of the POBA-based geopolymer improves the compressive strength significantly as it facilitates the formation of Al-O-Si bonds and speeds up the geopolymerization. Based on the experimental results, the optimum parameters for POBA-based geopolymers that yield the highest compressive strength are 15 M of NaOH, S/L mass ratio of 1.5, and 25% of Al_2O_3. As compared with run 2, the compressive strength of the POBA-based geopolymer in run 11 has increased up to 18.55 MPa at 7 days of curing after the addition of Al_2O_3. The high compressive strength of POBA-based geopolymer is attributed to the formation of more N-A-S-H gel within the geopolymer matrix, as indicated from the FTIR analysis. While these findings highlight the potential of POBA-based geopolymers as a sustainable and cost-effective alternative to OPC, it is important to consider environmental durability factors. For instance, challenges such as chloride ingression, carbonation of cement, and sulfate attack are common problems in construction that compromise the cement integrity. Therefore, future research should focus on assessing the resistance of POBA-based geopolymer to these degradation mechanisms to ensure their long-term viability in various environmental conditions.

Funding. This work was financial supported by Fundamental Research Grand Scheme (FRGS/1/2022/TK08/CURTIN/02/2) from Ministry of Higher Education (MOHE), Malaysia.

References

1. Habert, G.: Environmental impact of Portland cement production. in: Eco-Efficient Concrete, pp. 3–25. Elsevier Ltd., 2013. https://doi.org/10.1533/9780857098993.1.3
2. Amran, Y.H.M., Alyousef, R., Alabduljabbar, H., El-Zeadani, M.: Clean production and properties of geopolymer concrete; a review. J. Clean Prod. **251** (2020). https://doi.org/10.1016/j.jclepro.2019.119679
3. Adjei, S., Elkatatny, S., Aggrey, W.N., Abdelraouf, Y.: Geopolymer as the future oil-well cement: a review. J. Pet. Sci. Eng. **208** (2022). https://doi.org/10.1016/j.petrol.2021.109485
4. Tippayasam, C., Leonelli, C., Chaysuwan, D.: Effect of Agricultural Wastes With Fly Ash on Strength of Geopolymers, n.d
5. Khankhaje, E., et al.: On blended cement and geopolymer concretes containing palm oil fuel ash. Mater. Des. **89**, 385–398 (2016). https://doi.org/10.1016/j.matdes.2015.09.140
6. Hamada, H.M., Skariah Thomas, B., Tayeh, B., Yahaya, F.M., Muthusamy, K., Yang, J.: Use of oil palm shell as an aggregate in cement concrete: a review. Constr. Build. Mater. **265** (2020). https://doi.org/10.1016/j.conbuildmat.2020.120357
7. Yahya, Z., et al.: Chemical and Physical Characterization of Boiler Ash from Palm Oil Industry Waste for Geopolymer Composite, 2013. http://www.revistadechimie.ro1408
8. Castillo, H., Collado, H., Droguett, T., Vesely, M., Garrido, P., Palma, S.: State of the art of geopolymers: a review. E-Polymers **22**, 108–124 (2022). https://doi.org/10.1515/epoly-2022-0015
9. Singh, N.B., Middendorf, B.: Geopolymers as an alternative to Portland cement: an overview. Constr. Build. Mater. **237** (2020). https://doi.org/10.1016/j.conbuildmat.2019.117455
10. Salih, M.A., Abang Ali, A.A., Farzadnia, N.: Characterization of mechanical and microstructural properties of palm oil fuel ash geopolymer cement paste. Constr. Build. Mater. **65**, 592–603 (2014). https://doi.org/10.1016/j.conbuildmat.2014.05.031
11. Mijarsh, M.J.A., Megat Johari, M.A., Ahmad, Z.A.: Synthesis of geopolymer from large amounts of treated palm oil fuel ash: application of the Taguchi method in investigating the main parameters affecting compressive strength. Constr. Build. Mater. **52**, 473–481 (2014). https://doi.org/10.1016/j.conbuildmat.2013.11.039

12. Isa, M.N., Awang, H.: Characteristics of palm oil fuel ash geopolymer mortar activated with wood ash lye cured at ambient temperature. J. Build. Eng. **66** (2023). https://doi.org/10.1016/j.jobe.2023.105851
13. Ranjbar, N., Mehrali, M., Behnia, A., Alengaram, U.J., Jumaat, M.Z.: Compressive strength and microstructural analysis of fly ash/palm oil fuel ash based geopolymer mortar. Mater. Des. **59**, 532–539 (2014). https://doi.org/10.1016/j.matdes.2014.03.037
14. Hawa, A., Tonnayopas, D., Prachasaree, W.: Performance evaluation of metakaolin based geopolymer containing parawood ash and oil palm ash blends. Medziagotyra **20**, 339–344 (2014). https://doi.org/10.5755/j01.ms.20.3.4543
15. Alias Tudin, D.Z., Rizalman, A.N., Asrah, H.: Performance of palm oil fuel ash and rice husk ash based geopolymer mortar. In: E3S Web of Conferences, EDP Sciences, 2018. https://doi.org/10.1051/e3sconf/20186502011
16. Panias, D., Giannopoulou, I.P., Perraki, T.: Effect of synthesis parameters on the mechanical properties of fly ash-based geopolymers. Colloids Surf. A Physicochem. Eng. Asp. **301**, 246–254 (2007). https://doi.org/10.1016/j.colsurfa.2006.12.064
17. Kwek, S.Y., Awang, H., Cheah, C.B.: Influence of liquid-to-solid and alkaline activator (Sodium silicate to sodium hydroxide) ratios on fresh and hardened properties of alkali-activated palm oil fuel ash geopolymer. Materials **14** (2021). https://doi.org/10.3390/ma14154253
18. Benavent, V., et al.: Impact of aluminum on the structure of geopolymers from the early stages to consolidated material. Cem. Concr. Res. **90**, 27–35 (2016). https://doi.org/10.1016/j.cemconres.2016.09.009
19. Durak, U.: The improvement of strength and microstructural properties of fly ash-based geopolymer by adding elemental aluminum powder. J. Mater. Cycles Waste Manag. **25**, 157–170 (2023). https://doi.org/10.1007/s10163-022-01520-8
20. Jurowski, K., Grzeszczyk, S.: The influence of concrete composition on Young's modulus. Procedia Eng. 584–591. Elsevier Ltd. (2015). https://doi.org/10.1016/j.proeng.2015.06.181
21. Hardjito, D., Wallah, S.E.: Effect of Mixing Time and Rest Period on the Engineering Properties of Fly Ash-based Geopolymer Concrete, 2005. https://www.researchgate.net/publication/43649865
22. Li, G., Zhao, Y., Pang, S.S., Li, Y.: Effective young's modulus estimation of concrete. Cem. Concr. Res. **29**, 1455–1462 (1999). https://doi.org/10.1016/S0008-8846(99)00119-2
23. Yahya, Z., Abdullah, M.M.A.B., Hussin, K., Ismail, K.N., Razak, R.A., Sandu, A.V.: Effect of solids-to-liquids, Na2SiO3-to-NaOH and curing temperature on the palm oil boiler ash (Si + Ca) geopolymerisation system. Materials **8**, 2227–2242 (2015). https://doi.org/10.3390/ma8052227
24. Mijarsh, M.J.A., Megat Johari, M.A., Ahmad, Z.A.: Compressive strength of treated palm oil fuel ash based geopolymer mortar containing calcium hydroxide, aluminum hydroxide and silica fume as mineral additives. Cem. Concr. Compos. **60**, 65–81 (2015). https://doi.org/10.1016/j.cemconcomp.2015.02.007

Author Index

C. Yuan (Ed.): ICRER 2024, ESE, pp. 171–172, 2025.
https://doi.org/10.1007/978-3-031-93295-3

The manufacturer's authorised representative in the EU is Springer Nature Customer Service Centre GmbH, Europaplatz 3, 69115 Heidelberg, Germany. If you have any concerns regarding our products, please contact ProductSafety@springernature.com

Printed and bound by CPI Group (UK) Ltd, Croydon, CR0 4YY
30/09/2026
02225577-0017